Developing Mobile Web ArcGIS Applications

Learn to build your own engaging and immersive geographic applications with ArcGIS

Matthew Sheehan

BIRMINGHAM - MUMBAI

Developing Mobile Web ArcGIS Applications

First published: February 2015

Production reference: 1240215

Published by Packt Publishing Ltd.
Livery Place
35 Livery Street
Birmingham B3 2PB, UK.

ISBN 978-1-78439-579-7

www.packtpub.com

Credits

Author
Matthew Sheehan

Reviewers
Mark Cederholm
Andy Gup
Antti Kajanus
Noah Sager

Commissioning Editor
Kartikey Pandey

Acquisition Editor
Rebecca Youé

Content Development Editor
Siddhesh Salvi

Technical Editor
Vivek Pala

Copy Editor
Jasmine Nadar

Project Coordinator
Nidhi J. Joshi

Proofreaders
Lawrence A. Herman
Paul Hindle
Kevin McGowan

Indexer
Rekha Nair

Production Coordinator
Nilesh R. Mohite

Cover Work
Nilesh R. Mohite

About the Author

Matthew Sheehan is the founder and principal of WebMapSolutions (http://www.webmapsolutions.com/). He has over 20 years of experience working with both Esri and open source GIS technologies. Much of his current focus is on helping to expand the understanding and use of GIS. Using web technology, he is working with his technical team to build flexible ArcGIS solutions that can be integrated with existing enterprise systems. Disconnected use of GIS in the browser is a key part of this work.

Matthew has a bachelor's degree in geography from University of London and a master's degree in Applied GIS from the University of Utah.

About the Reviewers

Mark Cederholm, GISP, has over 20 years of experience in developing GIS applications using various Esri technologies, from Arc/Info AML to ArcObjects to ArcGIS Runtime and Web SDKs. He lives in Flagstaff, Arizona.

Andy Gup is a developer evangelist with Esri who focuses on building Mobile web and native Android map applications. He works on everything from small websites to enterprise systems, and he is also involved in a number of open source projects. You can check out his blog at http://www.andygup.net or reach him on Twitter at http://www.twitter.com/agup.

Antti Kajanus is a Finnish programmer who is very interested in mobile development and focuses on .NET client technologies. For the last 4 years, he has been creating customer solutions that are based on ArcGIS Server and ArcGIS Online. For the first 3 years, he was working as a software architect at Esri, Finland and after that he joined Esri's ArcGIS Runtime for the .NET development team. Currently, he lives in Edinburgh, Scotland.

He is an active presenter and loves to talk about ArcGIS application development and development practices. He's also a familiar face at Esri's Developer Summit conferences. If he's not building awesome GIS technology or applications, you can most likely find him roaming the Scottish mountains.

I would like to thank Packt Publishing and Matthew Sheehan for spreading the joy of ArcGIS development. I would also like to thank Nidhi Joshi for keeping the review process in control and for trying to keep me on schedule.

Noah Sager is a writer of code and prose and he is based out of a tiny mountain town in southern California. Originally from New Hampshire, he got a BSc in biology from McGill University and an MA in geography from Chicago State University. He recently received a GIS certificate from CSU and the TEFL Certificate from Literacy Works. He's worked in a variety of fields ranging from behavioral ecology to dairy farming, but it's GIS that he's truly passionate about. Currently, he supports GIS developers who are working with web and mobile applications and specialize in JavaScript, Flex, and iOS (Objective-C/Swift).

www.PacktPub.com

Support files, eBooks, discount offers, and more

For support files and downloads related to your book, please visit www.PacktPub.com.

Did you know that Packt offers eBook versions of every book published, with PDF and ePub files available? You can upgrade to the eBook version at www.PacktPub.com and as a print book customer, you are entitled to a discount on the eBook copy. Get in touch with us at service@packtpub.com for more details.

At www.PacktPub.com, you can also read a collection of free technical articles, sign up for a range of free newsletters and receive exclusive discounts and offers on Packt books and eBooks.

https://www2.packtpub.com/books/subscription/packtlib

Do you need instant solutions to your IT questions? PacktLib is Packt's online digital book library. Here, you can search, access, and read Packt's entire library of books.

Why subscribe?

- Fully searchable across every book published by Packt
- Copy and paste, print, and bookmark content
- On demand and accessible via a web browser

Free access for Packt account holders

If you have an account with Packt at www.PacktPub.com, you can use this to access PacktLib today and view 9 entirely free books. Simply use your login credentials for immediate access.

Dedicated to Cristi, Jack, and Lily.

Table of Contents

Preface

Mobile is transforming the GIS industry. Today, the demand from users of mobile devices for maps and location-based information is skyrocketing. ArcGIS is the most widely used GIS platform, with JavaScript as the most popular ArcGIS API. Esri's JavaScript team have been building a rich set of tools for developing ArcGIS Web applications.

This book will teach you how to build web-based mobile applications using the ArcGIS API for JavaScript.

What this book covers

Chapter 1, Introduction to Mobile Web ArcGIS Development, covers some of the core elements of developing mobile ArcGIS applications. We discuss how mobile web development is different from traditional web development. These differences include screen size, user interaction, design, functionality, and user and performance considerations. Mobile browsers and different development frameworks are also considered in this chapter.

Chapter 2, Understanding Mobile Frameworks and APIs, discusses some of the basic tools that are used for mobile web development; this includes popular frameworks and design paradigms. Dojo and jQuery mobile are both covered. In addition, simple code examples are included. We finish the chapter with an introduction to responsive, multi-device design and the Geolocation API.

Chapter 3, Building Your First Mobile ArcGIS Application, covers coding using the ArcGIS API. We introduce key tools, including Chrome's DevTools and the local web server installation. Starting with a brief review of basic coding elements, including HTML5, CSS3, and viewport, the bulk of this chapter walks you through code. We move from a very basic ArcGIS mobile application and evolve this code to include an ArcGIS map layer overlay and zooming to a users current location using geolocation.

Chapter 4, Advancing the Basic Mobile ArcGIS Application, is a natural progression from the previous chapter. We advance our mobile ArcGIS knowledge by adding more advanced functionality to our base application. This includes adding popular tools such as feature popups, a legend, finding features, and address search. At the completion of this chapter, you will have progressed from a basic to a more advanced understanding of ArcGIS mobile web development.

Chapter 5, Providing Cross-device Support with Responsive Design, Shows how to build mobile ArcGIS applications which run on all mobile devices. Here we introduce the idea of responsive design using the popular Bootstrap framework. We start with some of the key concepts, then start coding. We not only walk through the development of cross-device ArcGIS web applications with Bootstrap, but also evolve the base application to include a number of popular tools.

Chapter 6, Integration with ArcGIS Online, will focus on ArcGIS Online webmaps. So far, we worked with individual ArcGIS published services. Here, we will build a mobile ArcGIS application that works directly with ArcGIS Online. The application that we evolve in this chapter will include authentication and a webmap list and it will load a user-selected webmap.

Chapter 7, Developing Hybrid ArcGIS Mobile Applications with PhoneGap, demonstrates the flexibility of building mobile applications with the ArcGIS JavaScript API. We can not only build mobile applications that run in browsers, but also convert these applications to hybrid or installable apps that can be distributed to the various mobile app stores. In this chapter, we will provide guidance on working with PhoneGap and Cordova. We will not only go through setup, but also build a basic ArcGIS hybrid mobile application.

What you need for this book

To complete the exercises in this book, you will need access to a web browser, preferably Google Chrome. A text editor such as the free NotePad++ will be required to work with the many code samples. We recommend that you test these code samples using a locally running web server such as Apache or Internet Information Server(IIS). The majority of examples will access the publicly available instances of ArcGIS Server, so it will not be necessary for you to install the ArcGIS Server. You will need a free developer ArcGIS Online account for the ArcGIS Online discussion in *Chapter 6, Integration with ArcGIS Online.*

Who this book is for

This book targets GIS professionals who would like to learn how to create mobile web ArcGIS applications. It is primarily oriented towards beginners and intermediate-level GIS developers, and application developers who are interested in, or have been tasked with, implementing mobile Web ArcGIS solutions. The focus of the book will be on the ArcGIS API for JavaScript; for this, prior experience will be helpful but it is not required. Mobile applications will be built for both ArcGIS Server and ArcGIS Online; again, no prior experience is required for this.

Conventions

In this book, you will find a number of styles of text that distinguish between different kinds of information. Here are some examples of these styles, and an explanation of their meaning.

Code words in text, database table names, folder names, filenames, file extensions, pathnames, dummy URLs, user input, and Twitter handles are shown as follows: "Mobile developers will need to use require() to load any of the additional modules."

A block of code is set as follows:

```
<script>
  require(["esri/map", "dojo/domReady!"],
    function(Map){
    });
</script>
```

When we wish to draw your attention to a particular part of a code block, the relevant lines or items are set in bold:

```
if (navigator.geolocation) {
    navigator.geolocation.getCurrentPosition(showPosition);
} else {
    x.innerHTML = "Geolocation is not supported by this browser.";
}
```

Any command-line input or output is written as follows:

```
>npm install cordova -save
```

New terms and **important words** are shown in bold. Words that you see on the screen, in menus or dialog boxes for example, appear in the text like this: "When you click on the sign-in button, it redirects you to the OAuth Log in page."

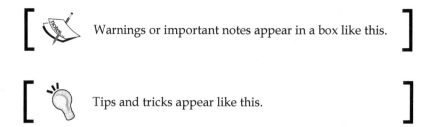

Warnings or important notes appear in a box like this.

Tips and tricks appear like this.

Reader feedback

Feedback from our readers is always welcome. Let us know what you think about this book—what you liked or may have disliked. Reader feedback is important for us to develop titles that you really get the most out of.

To send us general feedback, simply send an e-mail to feedback@packtpub.com, and mention the book title via the subject of your message.

If there is a topic that you have expertise in and you are interested in either writing or contributing to a book, see our author guide on www.packtpub.com/authors.

Customer support

Now that you are the proud owner of a Packt book, we have a number of things to help you to get the most from your purchase.

Downloading the example code

You can download the example code files for all Packt books you have purchased from your account at http://www.packtpub.com. If you purchased this book elsewhere, you can visit http://www.packtpub.com/support and register to have the files e-mailed directly to you.

Errata

Although we have taken every care to ensure the accuracy of our content, mistakes do happen. If you find a mistake in one of our books — maybe a mistake in the text or the code — we would be grateful if you would report this to us. By doing so, you can save other readers from frustration and help us improve subsequent versions of this book. If you find any errata, please report them by visiting http://www.packtpub.com/submit-errata, selecting your book, clicking on the **errata submission form** link, and entering the details of your errata. Once your errata are verified, your submission will be accepted and the errata will be uploaded on our website, or added to any list of existing errata, under the Errata section of that title. Any existing errata can be viewed by selecting your title from http://www.packtpub.com/support.

Piracy

Piracy of copyright material on the Internet is an ongoing problem across all media. At Packt, we take the protection of our copyright and licenses very seriously. If you come across any illegal copies of our works, in any form, on the Internet, please provide us with the location address or website name immediately so that we can pursue a remedy.

Please contact us at copyright@packtpub.com with a link to the suspected pirated material.

We appreciate your help in protecting our authors, and our ability to bring you valuable content.

Questions

You can contact us at questions@packtpub.com if you are having a problem with any aspect of the book, and we will do our best to address it.

1
Introduction to Mobile Web ArcGIS Development

We live in interesting times. Mobile and cloud computing are changing how and where we are able to use technology. No longer are we limited to our home or office. Today, technology is accessible and usable at any location and at anytime. By technology, we mean computers, smartphones, tablets, and the new phablets.

Mobility means the ability to change location. This has suddenly made location and location-based data and analysis terribly important. **Geographic Information Systems (GIS)** is a location-focused technology. It provides the means to collect, store, visualize, and analyze any and all location-based data.

Today's cloud-based mobile world heralds the beginnings of a revolution in location technology and GIS.

This book will introduce you to mobile application development using Esri's mobile ArcGIS JavaScript API; it presumes that you are familiar with HTML, JavaScript, and CSS. The API provides a rich set of tools for developing high performance, engaging mobile ArcGIS applications. You will learn about the classes available in the API for mobile development, how to use them in a JavaScript-based mobile web application, and how to incorporate ArcGIS services and ArcGIS Online content to enhance your applications. ArcGIS RESTful web services will also be introduced, which will provide resources via argument/value pairs. The ArcGIS Feature service, that will be introduced early in the book, is one of the many examples. In this chapter, we will go through some of the fundamentals: how mobile development is different from standard web development and design and performance considerations, mobile device types and popular mobile frameworks that are now available, and the more advanced changes that the rapid advancement in mobile- and cloud-based technologies are bringing to the GIS industry.

The topics covered in this chapter are as follows:

- Web ArcGIS development
- Differences between mobile development and traditional web development
- Introduction to mobile frameworks
- How mobile web is different from traditional web
- The impact of mobile and cloud-based technologies
- The Cloud on the GIS industry

Fundamentally, Web ArcGIS development is focused on providing users access, through any web browser, to services provided by the ArcGIS platform: ArcGIS Server, Portal for ArcGIS, and ArcGIS Online. Portal for ArcGIS provides the same experience as ArcGIS Online but within an organization's infrastructure (on-premise or in the cloud). This is a particularly good solution where there are concerns around security. The following screenshot illustrates the mobile ArcGIS application running on an iPad:

Mobile ArcGIS application running on an iPad

Often, mobile ArcGIS application begins with an interactive map that consists of a base map overlaid with point, line, or polygon layers. Click or tap on one of these features and a pop should appear listing the feature's attributes.

ArcGIS web applications are not only used for visualization, but they often provide specific functionality too. Tools can be built that give users specific workflows. These tools might include search, identification, buffering, measurement, and more. One of the attractions of ArcGIS technology is the number of RESTful web services that are provided. There is an expanding list of services such as those that transform an address to a location (geocoding), a location's surrounding area demographics (geoenrichment), and projection (geometry). Web services provide the base for the tools that developers build into their mobile web ArcGIS applications. ArcGIS is a service-rich platform that can be easily accessed through one of the many web APIs such as the ArcGIS JavaScript API.

The transition from standard web development to mobile development needs careful planning and consideration. Mobile web development is different. It is generally targeted at a multitude of low-powered mobile computers that are limited in memory, storage, and processing power. Combine this with the range of different mobile device types and screen sizes; design and performance are key considerations when planning the development of any mobile ArcGIS web application.

It is worth discussing some of the differences between standard and mobile web development in a little more detail. There are some key areas that should be mentioned, notably the screen size, user interaction, design, functionality, and performance considerations.

Downloading the example code

You can download the example code files for all Packt books you have purchased from your account at http://www.packtpub.com. If you purchased this book elsewhere, you can visit http://www.packtpub.com/support and register to have the files e-mailed directly to you.

Screen size

There is a wide range of mobile device types, from smartphones to phablets and tablets. Screen sizes vary widely across these different devices, ranging from 3.5 inch smartphones to 10.1 inch tablets. Screen size affects how users interact with mobile web applications. Pixel density also varies between devices. For example, map symbology might look good on an older Android tablet but small on an iPad with a retina screen. The following screen shot illustrates different screen sizes:

Interacting with mobile applications

By interaction, we mean how users work with an application, which includes the following:

- Change from mouse to finger interaction, that is, from click to tap
- New mobile data input methods including popup keyboards
- The need to provide user interaction feedback

Next, we will cover these interaction differences in a little more detail.

From clickable to tappable

The following screenshot illustrates the Smartphone ArcGIS finger interactions that we just discussed:

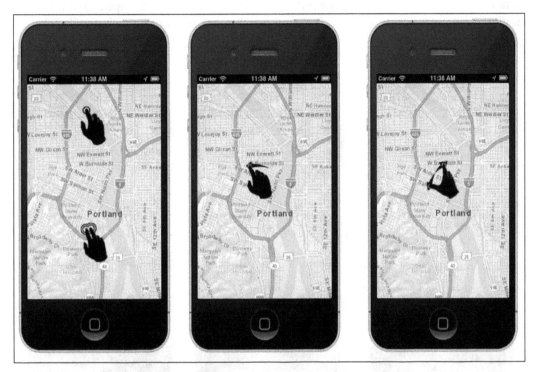

Smartphone ArcGIS finger interaction

Traditional web development is focused on mouse interaction, where a mouse is a high-precision device. Mobile web development is quite different, as it relies on touch or finger interaction instead. This is very low precision, particularly since we all have different sized fingers. Tapping a button can be problematic and requires that the selectable on-screen items be larger to ensure a good user experience. Map interaction is driven by a finger slide for pan and a pinch to zoom. Zoom sliders are often included in mobile ArcGIS apps and they also provide zoom functionality.

New data input and collection methods

Data input relies on a screen-based, touch-driven keyboard. Usually, multiple keyboards are available; these are for character input, phone or numeric input, and date input, respectively. Voice is another potential source of data input for a mobile device. This provides new and interesting possibilities for mobile web developers.

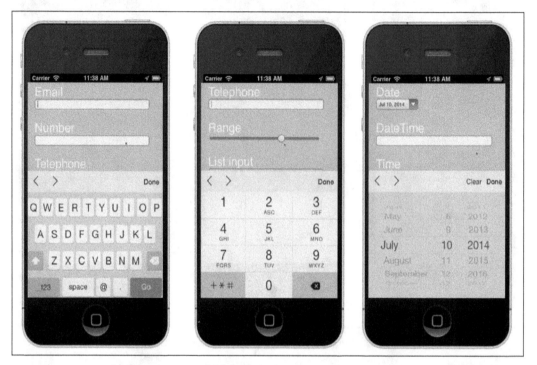

Mobile device input types

For ArcGIS developers, the built-in GPS that is present on most devices means that current location, or geolocation, is a valuable new data source for use in any mobile application. Similarly, built-in cameras provide both video and still imagery, and these can help to enrich the data used or collected in an application.

Providing interaction feedback

Any interaction should provide users with obvious feedback. When a user taps a button or link, it is good if the item changes state.

 Feedback is particularly important to help guide mobile users, when they use your application.

For example, as shown in the following screenshots, a green-colored button with the label, 'Online', changes to red and the label changes to 'Offline' after a user tap:

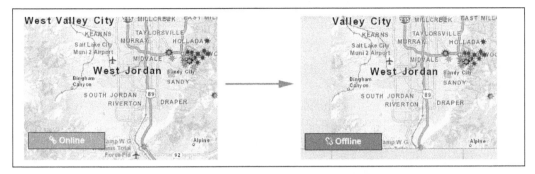

Button changes color and label based on user interaction

Another good feedback mechanism is a loading animation. This is particularly helpful when loading maps:

Map loading animation

Designing for mobile

Careful consideration is needed when designing mobile ArcGIS applications. Key areas to consider include the following:

- Mobile web applications need to be simple and easy to use
- Given the range of mobile screen sizes and pixel densities, layout design becomes important
- Orientation of a mobile device after it is rotated, as this will directly affect your mobile application and the layout of user interface components
- Differences between mobile device brands such as Apple, Google, Windows, and Blackberry
- The target audience for both the application and the device
- Styling differences between traditional and mobile web applications

Again, let's consider each of these design differences in a little more detail.

Simplicity and intuition

Too often mobile web developers try to recreate a traditional web experience which results in an application that is overly complex. Simplicity and intuition are key. Don't try to do all things for all people. Workflows need to be obvious so that if a user selects a menu item or taps a button, then what they see on the screen makes sense. One screen should logically follow the next.

A GIS map is made up of a base map with so-called feature overlays. These overlays are the map layers and represent geographic features through a point, line, or polygon. Points might represent water valves, lines might be buried pipelines, while polygons might represent parks.

Navigation should be simple. For example, if a user has tapped on a point feature and is viewing a full screen attribute inspector, then returning to the map should be easy and obvious.

Focus on designing your mobile web apps so that they are simple and intuitive. Mobile users will stop using a mobile application if it has a cluttered interface and is hard to use.

Layout

Careful consideration needs to go into mobile ArcGIS app layouts. Single column layouts combined with collapsible menus often work best. Traditional ArcGIS web applications are often designed with multiple columns, commonly with a layer list on the left-hand side, the map at the center, and a list of tools on the right-hand side.

The following screenshot illustrates the multiple versus single column layouts that we just discussed:

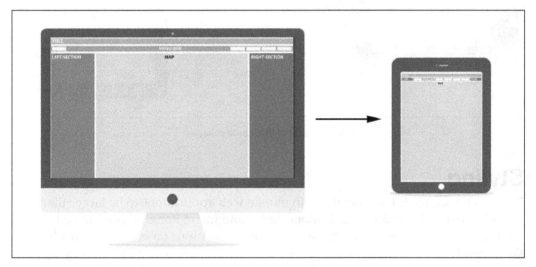

Multiple versus single column layouts

Collapsable menus provide ways to easily combine simple interfaces with functional options: widgets and tools. Using a responsive design, as we will discuss in *Chapter 5, Providing Cross-device Support with Responsive Design*, allows a multi-column layout to be adapted into a single-column layout depending on the screen size of the mobile device that is accessing the application.

Orientation

By rotating their mobile device, users can change the orientation of an application to either landscape or portrait. An orientation change affects layout. For example, a map that has a column on the right-hand side might work well in the landscape mode, but it might obscure the map if a user rotates the device to the portrait mode.

The following screenshot illustrates the mobile device orientation change:

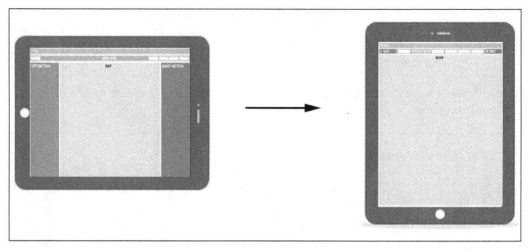

Mobile device orientation change

Styling

Finger tap requires interactive items in mobile web applications to be larger than in traditional web applications. Buttons, links, and menu items all need to be larger than their traditional web counterparts. Recommendations on size vary considerably. For example, the recommended size for buttons varies from 44 x 44 pixels by Apple for iPhones to 28 x 28 pixels as per Nokia's smartphone suggestion. Either way, on-screen items on a mobile device need to be easy to select.

We will discuss style sheets or CSS in depth in later chapters. These greatly help developers style their mobile ArcGIS apps. Both Android and Apple provide style guides and these can be found at the following links:

- `http://developer.android.com/design/style/index.html`
- `https://developer.apple.com/library/ios/documentation/UserExperience/Conceptual/MobileHIG/index.html`

By contrast, some on-screen items are smaller on mobile devices. Popups are a good example of this.

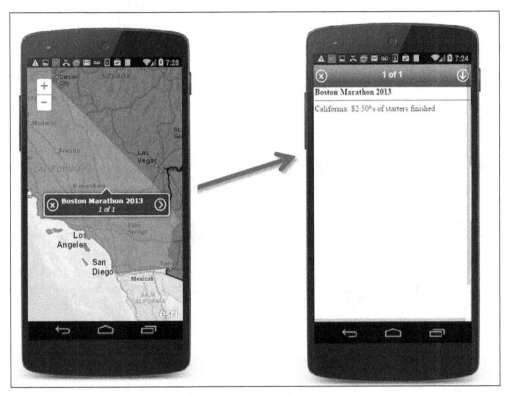

Small popups open to attributes screen

Mobile apps are often styled for a specific device. Apple and Android have styling and design guidelines. These are recommendations for the look and feel of mobile apps on individual platforms. Though it is a little more difficult to style a mobile application for a specific platform using web technology than it is for native mobile applications, much can be done now with style sheets to provide a platform-specific look and feel to your mobile web application by targeting mobile-specific functionality.

Mobile applications are often used for different purposes than traditional web applications. Their requirements and workflows are quite different from to desktop browser applications. Thus, functionality is a key consideration. This includes the following:

- Building a mobile application on a mobile device that makes sense from a functional perspective.
- Avoid mirroring, recreating, or reusing a traditional web application for mobile use.

- Thinking carefully about your mobile ArcGIS users. Are they GIS or non-GIS users? Functionality will differ depending on the audience. We no longer build web applications that provide tools for everybody (we hope!).

Mobile targeted functionality

There is usually a clear divide between what functionality makes sense for mobile use versus desktop use. Questions around current location or use of GPS are usually mobile use cases. More complex, analytical tasks are more likely to be part of the desktop world.

Focused mobile applications

When web ArcGIS applications started to become popular, too often developers tried to recreate desktop applications for the Web. This resulted in complex applications loaded with tools that were cluttered and in many ways unusable. Web GIS had a distinctly narrower focus than desktop GIS. As we move to mobile ArcGIS development, developers need to remember lessons that they learned from the early days of the Web. Mobile web GIS again has a narrower focus than traditional web GIS. A good example of a focused mobile ArcGIS app might be one that just provides editing capabilities: add, edit, or delete a point feature. In the mobile world, focused functionality is king.

Target audience GIS versus non-GIS users

You need to build your mobile ArcGIS application for your intended audience. If your app is for maintenance workers who need to report on their work of filling pot holes, you will need to build an app that includes all the functional elements—a map with relevant layers, reporting tool, and so on—that are required to complete this task. GIS has traditionally served a GIS-trained audience. As we will discuss later, the combination of mobile and cloud technology has widened the potential user base for applications powered by ArcGIS.

 The world of GIS is changing. As the technology becomes more popular, the number of users unfamiliar with GIS is increasing dramatically. Developers are now building seeing it for the first time GIS-powered web applications for a new audience.

Today, a new group of users is accessing and using services provided by ArcGIS. Functionality, design, and workflows need to be carefully considered for this new audience.

Fast responding mobile applications

Though this is not an exhaustive list, the last key difference that we will discuss between traditional and mobile web development is performance.

Users expect mobile applications to be fast and responsive. Think about how people use their desktops or smartphones today. Gaming and social media are extremely popular. High performance is an expected factor. We will be discussing performance and best practices throughout the book

One area where mobile web is criticized when compared with native mobile development is in the area of performance. Extra effort and attention is needed during mobile web development to optimize performance. This is also true for native apps and backend services.

Network issues potentially plague all mobile applications. Lack of or poor Wi-Fi connectivity can cause any mobile application to be either slow or unusable. Though beyond the scope of this book, advances in offline or disconnected ArcGIS mobile web application are helping to reduce the impact of network issues.

Working with mobile browsers

There are an increasing number of mobile browsers that are now available. Internet Explorer and Safari are the default browsers for Windows and iOS, respectively. Other browsers including Opera and Dolphin are also becoming more popular. Similar to traditional web development, cross-browser compatibility needs to be carefully considered. This is because mobile functionality that works well in one browser might not work quite so well in a different browser. Testing across all popular browsers remains an important part of mobile web development.

> Android devices are shipped with a variety of default browsers. These are often described generically as the 'Android browser'. This can present challenges that are described in a presentation that can be found at `http://slides.com/html5test/the-android-browser#/`.

There are an increasing number of resources such as modernizer, yepnope, and `caniuse.com` to help you with the challenges of testing.

> See the excellent article in Mashable on mobile testing tools. The article can be found at: `http://mashable.com/2014/02/26/browser-testing-tools/ #`.

Web, native, and hybrid mobile applications

There are three main types of mobile applications: web, native, and hybrid. Once, native was the preferred approach for mobile development, but web technology is advancing at a rapid rate and, in many situations, it can be argued to be there is actually a better choice than native. Flexibility is a key benefit of mobile web; one code base is all that is needed for an app to run across different devices and platforms. In the native world, the same mobile application that runs on iOS, Android, and Windows platforms requires three separate development teams. These developers would need to write the code in Objective-C for iOS, in Java for Android, and in .NET for Windows.

In *Chapter 7, Developing Hybrid ArcGIS Mobile Applications with PhoneGap* we will discuss hybrid apps. These are mobile web applications that can be converted, using technologies such as Cordova/PhoneGap, into installable apps that more closely resemble their native counterparts, again demonstrating the flexibility of mobile web development.

Mobile frameworks, toolkits, and libraries

JavaScript is an implementation of the **ECMAScript** language open standard. The JavaScript community is worldwide and very active. JavaScript is one of the most widely used languages. Today, there are a plethora of tools that are provided by this community that help JavaScript development. These tools go under different names: toolkits, libraries, modules, and frameworks.

 JavaScript frameworks, toolkits, and libraries help to simplify mobile web development.

Often the terms library and framework are used interchangeably. There is a subtle but distinct difference between a library and framework called the inversion of control. When a developer's code calls a library, the code is in control. The reverse is true for a framework. Here, the framework calls the developer's code. More simply put, a library is a collection of functionality that you can call, whereas a framework provides automatic flow controls. For all intents and purposes, libraries and toolkits are identical.

In later chapters, both frameworks and libraries will be discussed in more detail. The code examples provided in the code bundle will be built using the Dojo framework, this is the base for the ArcGIS JavaScript API. Dojo helps developers to build dynamic web interfaces.

An alternative to Dojo is jQuery mobile.

There are two other frameworks that are worth mentioning and will form the base for later chapters. These are Bootstrap and PhoneGap.

Bootstrap

Bootstrap is the most popular framework for developing responsive mobile applications. This framework provides automatic layout adaptation. That includes adapting to changes in device orientation and screen size. This means your ArcGIS web application will look good and be usable on all mobile devices: smartphones, phablets, and tablets.

The following screenshot shows Bootstrap downloads, illustrating its popularity:

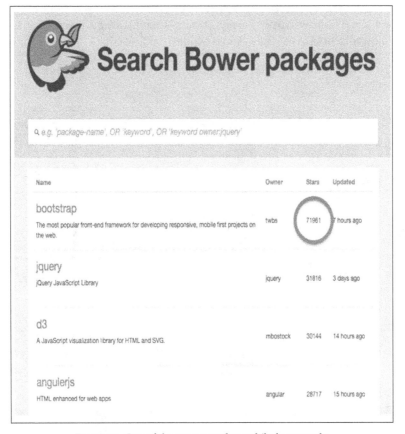

Bootstrap: One of the most popular mobile frameworks

PhoneGap

PhoneGap is a framework that allows developers to convert web mobile applications into installable hybrid apps.

> Cordova is an open source framework that is managed by the Apache Foundation. PhoneGap is based on Cordova and PhoneGap is owned and managed by Adobe.
>
> For more information, visit `http://cordova.apache.org/`.
>
> The names (Cordova and PhoneGap) are often used interchangeably; in fact, the two are very similar, but there is a legal and technical difference.

These hybrid apps can be distributed through the various mobile app stores, just like their native counterparts. In *Chapter 7, Developing Hybrid ArcGIS Mobile Applications with PhoneGap* we will walk you through the development of a hybrid ArcGIS mobile app. PhoneGap works by providing a web app with a native wrapper around a headless browser. These wrappers are platform-specific and allow your web app to be built to run natively on any of the popular mobile platforms.

The following screenshot illustrates how to convert a web app to hybrid using PhoneGap:

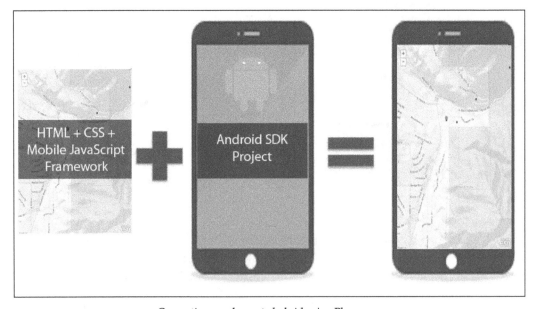

Converting a web app to hybrid using Phonegap

Summary

Mobile ArcGIS web development is an exciting and ever more popular way to build mobile applications. Though the technology is the same as that used for traditional website development, there are many additional areas that developers need to consider when embarking on a mobile web project. These include different screen sizes, user interaction, design, functionality, and user and performance considerations. Flexibility is one of the key advantages of mobile web development. It is now possible to build web applications that can be used on all popular mobile devices and platforms.

In the next chapter, we will consider some of the core building blocks for developing a mobile ArcGIS web application. These are some of the frameworks and APIs that will become very familiar to you and will help you to advance your mobile ArcGIS development skills.

2
Understanding Mobile Frameworks and APIs

In this chapter, we will discuss some of the basic ideas behind mobile web development: popular frameworks, design paradigms, and the Geolocation API.

The frameworks and APIs discussed in this chapter will form the base for the code samples and discussions in later chapters. The topics covered in this chapter are as follows:

- Esri's ArcGIS JavaScript API
- Dojo and jQuery Mobile
- Bootstrap and Responsive design
- Geolocation API

Esri ArcGIS JavaScript API

The ArcGIS JavaScript API provides a rich set of tools to build mobile ArcGIS apps. As we will discuss later, it is built on the Dojo framework. The API can be used both for mobile and desktop web application development. In this chapter, we will introduce Dojo and other frameworks, APIs, and libraries relevant to ArcGIS JavaScript development.

Esri describes the ArcGIS API for JavaScript at
`https://developers.arcgis.com/javascript/jshelp/`
as follows:

"The ArcGIS API for JavaScript is a lightweight way to embed
maps and tasks in web applications. You can get these maps
from ArcGIS Online, your own ArcGIS Server or others' servers."

The API documents can be found at the website.

Dojo and jQuery mobile

In *Chapter 1, Introduction to Mobile Web ArcGIS Development*, we discussed the
difference between frameworks, libraries, and toolkits. These terms are somewhat
subjective. At the end of the day, these are all built on JavaScript. They are simply
structures and tools that make development easier and faster.

Mobile development conversations in relation to these structures and tools often
revolve around the use of Dojo and jQuery Mobile. It's worth spending a little time
discussing both Dojo and jQuery.

The ArcGIS API has been built using Dojo. However, as we
will show later, jQuery Mobile can still be used with the API.

Dojo

Dojo is one of the many open source JavaScript frameworks that is used for
constructing dynamic web user interfaces. At its core, it is a simple collection of
JavaScript files. One line in your code referencing dojo.js is all you need to access
these files:

```
<script src="//ajax.googleapis.com/ajax/libs/dojo/1.9.4/dojo/dojo.js"
        data-dojo-config="async: true"></script>
```

This simple reference takes care of the initialization of Dojo.

Normally, once a JavaScript library file is loaded into the browser, all its methods,
properties, and classes are available.

Dojo recently adopted **the Asynchronous Module Definition (AMD)** format for its source code, allowing for modular web development. You will often see reference to AMD in Dojo documentation.

 Dojo modules allow developers to load specific code chunks and to break their own code into discrete blocks.

JavaScript code modules are written in AMD. A module can be accessed using a single reference. Modules are helpful when developers want to split their code into logical subsets to handle a specific functionality. They are also reusable since modules can be added by any new or existing application. If you want to represent a feature with information like type and perhaps add some methods to your feature, it makes sense to put all that code in a single location.

With dojo.js you will get an AMD loader that defines two global functions: `require` loads modules and `define` allows you to define your own modules. Typically, a module is a single JavaScript file.

 The Dojo loader includes two APIs: the AMD API and the legacy Dojo API. The AMD API allows modules to be loaded asynchronously and this decreases page load times. It also provides multiple platform support. You can find more information at http://dojotoolkit.org/reference-guide/1.8/loader/.

It's time to look at some code. Let's use a couple of basic Dojo modules: `dojo/dom` and `dojo/dom`-construct:

```
<!DOCTYPE html>
<html>
<head>
    <meta charset="utf-8">
    <title>Dojo Hello</title>
</head>
<body>
    <h1 id="greeting">Dojo</h1>
    <!-- load Dojo -->
    <script src="//ajax.googleapis.com/ajax/libs/dojo/1.9.4/dojo/dojo.
js"
            data-dojo-config="async: true"></script>
    <script>
        require([
            'dojo/dom',
```

```
                        'dojo/dom-construct'
        ], function (dom, domConstruct) {
                var greetingNode = dom.byId('greeting');
                domConstruct.place('<i> Hello!</i>', greetingNode);
        });
    </script>
</body>
</html>
```

For browser-based JavaScript, the **Document Object Model (DOM)** is the canvas on which we paint the user interface. Once loaded, we can manipulate this content through the DOM. Dojo helps to make working with the DOM easier. The `dojo/dom` module defines the core Dojo DOM API. This provides DOM management and manipulation. The `dojo/dom-construct` defines the core dojo DOM construction API.

These are important and highly useful modules. In the preceding example, we used `dojo/dom` to find a page element by ID and `dojo/dom-construct` to place a new element on the page. You can read more in the Dojo documentation at `http://dojotoolkit.org/documentation/tutorials/1.7/dom_functions/`.

Walking through the code, you will see that the first parameter, `require`, is an array of identifiers or IDs for the modules that you want to load; in our case, they are `dojo/dom` and `dojo/dom`-construct. This parameter maps directly to the file names contained in the source distribution of Dojo.

Pathnames to the Dojo modules actually represent the pathnames in the Dojo source directory structure.

The second parameter is a function. AMD loaders operate asynchronously. In JavaScript, these asynchronous operations are implemented with callbacks. This second parameter is thus a callback function.

A callback is a piece of executable code that is passed as an argument to another code.

The modules are passed by the AMD loader as parameters to the callback function (you will need to ensure that they are in the same order in which they were specified in the module ID array). There is no rule for naming the parameters for your callback function, but for the sake of consistency and readability, it is usually best to use names that are based on the module ID.

Some vendors such as Esri provide optional argument aliases for consistency. Here's a link to Esri's JS API argument aliases: `https://developers.arcgis.com/javascript/jsapi/argument_aliases.html`

In our preceding code example, inside the callback function, you will see that the modules are being used to get a reference to the DOM node and to manipulate its content:

```
var greetingNode = dom.byId('greeting');
```

Dojo Widgets and Plugins

Let's discuss some key elements of Dojo, namely widgets and plugins.

The following screenshot illustrates Dojo libraries and toolkits:

Dojo libraries and toolkits

Plugins are a special type of module. They are relatively new and extend the AMD loader to allow the loading of new features. Plugins are actually loaded in a similar way to modules. The syntax is slightly different a ! is used at the end of the identifier to indicate that this is a plugin request, as shown in the following piece of code:

```
define(["dojo/dom", "dojo/domReady!"], function(dom){
    //This function does not execute until the DOM is
//ready
    dom.byId("someElement");
});
```

Four of the more important and useful plugins are dojo/text, dojo/i18n, dojo/has, and dojo/domReady.

Dijit is a widget system that sits on top of Dojo. Widgets are UI objects that have layout and properties. They are HTML- and CSS-bound by JavaScript. Widgets are powerful GUI elements that require little or no JavaScript.

 One of Dojo's biggest advantages over other JavaScript frameworks is its Dijit UI framework.

Dijits can be implemented in one of two ways: the declarative approach and the programmatic approach. With the declarative approach, developers use special attributes inside regular HTML tags, as shown in the following code snippet:

```
<script>
require(["dojo/parser", "dijit/form/Button"]);
</script>
<button id="clickBtn" data-dojo-type="dijit/form/Button"
type="button">Click me first!
    <script type="dojo/on" data-dojo-event="click" data-dojo-
args="evt">
        require(["dojo/dom"], function(dom){
            dom.byId("result1").innerHTML += "Thanks! ";
        });
    </script>
</button>
<div id="result1"></div>
```

With the second or programmatic approach, widgets are directly used through JavaScript, as you can see in the following code snippet:

```
<script>
require(["dijit/form/Button", "dojo/dom", "dojo/domReady!"],
function(Button, dom){
    // Create a button programmatically:
```

```
    var myButton = new Button({
    id: "clickBtn",
        label: "Click me second!"
    }, "progButton");
});
    myButton.on("click", function(){dom.byId("result2").innerHTML +=
"Thanks!";});
    myButton.startup();
</script>
<button id="progButton" type="button"></button>
<div id="result2"></div>
```

 You can see the full code sample using the programmatic approach at:

`http://webmapsolutions.com/book/mobilearcgis/chapter2/programmaticsample.html`.

Dojo provides a high level of flexibility through its AMD modules, plugins, and the Dijit toolkit. It is also extensible and comes with a wide array of tools. In the later chapters, we will discuss Dojo and its use for mobile ArcGIS development in more depth.

jQuery Mobile

jQuery Mobile is an alternative to Dojo to build mobile ArcGIS apps.

 If you consider building an ArcGIS mobile app with jQuery mobile, you will still require Dojo to interact with the ArcGIS API for JavaScript

It provides many of the same benefits as Dojo. Just like Dojo, a single line takes care of initialization:

```
<script type="text/javascript" src="https://ajax.microsoft.com/ajax/
jQuery/jquery-1.4.2.min.js"></script>
```

jQuery syntax is, however, quite different from Dojo. For example, a dollar sign ($) denotes a jQuery constructor and the selector that follows is enclosed in parentheses. A selector can be an element (tag), an element ID, or an element class name. For example:

```
$("#jqSlider").slider("value", map.getLevel());
```

This code snippet references a slider element by its ID. The slider will be set to the value of the current map level.

ArcGIS web applications can be built with jQuery, but there is a requirement to use elements of Dojo. There is no jQuery-based ArcGIS API, so this might be more accurately called a "jQuery plus ArcGIS implementation". The following code snippet gives you the feel of what a jQuery ArcGIS implementation looks like:

```
<script>
  require(["esri/map", "dojo/on", "dojo/domReady!"],
  function (Map, on) {
    var map;
        map = new Map("map", {
        basemap: "streets",
        center: [2.352, 48.87],
        zoom: 12,
        slider: false
  });
  //Map load event
    map.on("load", function () {
// Hook up jQuery
      $(document).ready(jQueryReady);
  });
// jQuery Integration
    function jQueryReady () {
// Create jQuery Slider
          createSlider();
        }
    function createSlider () {
    $("#jqSlider").slider({
      min: 0,
      max: map.getLayer(map.layerIds[0]).tileInfo.lods.length - 1,
      value: map.getLevel(),
      orientation: "vertical",
      range: "min",
      change: function (event, ui) {
        map.setLevel(ui.value);
  }
  });
```

//Each time the map is zoomed, the slider default value is reset.

```
map.on("zoomEnd", function () {
 $("#jqSlider").slider("value", map.getLevel());
 });
 }
 });
</script>
<div id="jqSlider" style="position: absolute; right: 20px; top: 40px;
height: 180px; z-index: 2; font-size: 9px;"></div>
```

 To help developers get started, Esri has provided a downloadable jQuery sample in Github at https://github.com/Esri/jquery-mobile-map-js.

Dojo and Esri's ArcGIS API for JavaScript

In general, it makes sense for developers to take advantage of Dojo, jQuery Mobile, or the equivalent JavaScript tools. Normalizing browser inconsistencies, DOM APIs, event management tools, and much more are provided by these frameworks. Esri has chosen Dojo in favor of jQuery for their mapping APIs. This makes for a strong argument to use Dojo in your mobile ArcGIS development.

 The code examples in this book use Dojo. More information about Dojo and the GFX library can be found at https://dojotoolkit.org/documentation/.

Why did Esri choose Dojo? According to one of their senior engineers, because:

 "Dojo provides a more extensive set of features and tools, has a mature class/module system baked into the API as well as a robust widget lifecycle system. There are many things attractive about Dojo, including support for vector graphics via dojox.gfx."

Esri's official reasons for choosing Dojo as the foundation for the ArcGIS API for JavaScript include the following:

- The dojo build system
- It has an easy to use class-based inheritance system
- It has an AMD module loader for managing code in large applications
- It has a vector graphics abstraction via dojox/gfx to simplify drawing graphics using **Scalable Vector Graphics (SVG)**, Canvas, or **Vector Markup Language(VML)**
- Dojo and Dijit are both fully accessible
- Dojo fully supports internationalization

Dojo and jQuery Mobile form the foundation for many mobile web apps. Next, let's consider two other popular JavaScript toolsets that will greatly help you to develop stunning mobile ArcGIS apps.

> Esri uses Dojo to create the ArcGIS API for JavaScript. Developers only need to use Dojo to interact with Esri's JS API. Everything else can be done with pure JavaScript or other user interface frameworks. For UX and any other functionality, developers can choose whichever JS library that they wish to use.

Bootstrap and Responsive design

As mentioned in *Chapter 1, Introduction to Mobile Web ArcGIS Development*, Bootstrap is a very popular framework for building web interfaces. It is made up of a combination of JavaScript, CSS, and fonts. It supports both CSS3 and HTML5. The framework considers both device limitations and user behavior. Key to the popularity of Bootstrap is its responsive system. This allows developers to build a single mobile web app that works well across a variety of devices with different screen sizes. This means a web app that looks good on the largest monitor down to the smallest smartphone.

The following screenshot illustrates the Responsive design of a web app across varied screen resolutions:

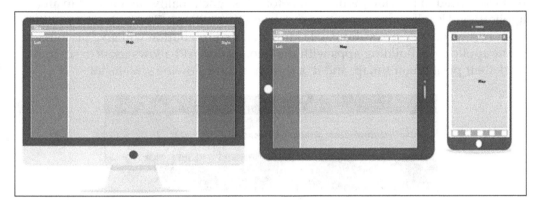

Responsive Design – One code base for all devices

Bootstrap implements a fluid grid system. This breaks the screen into columns and uses percentages instead of pixels for column widths. A mobile ArcGIS app built using Bootstrap adapts to different screen sizes using this fluid column system. The fluid grid system provides 12 columns for wider screens; these columns then stack as the screen size decreases. Thus, the same application might appear with multiple columns on an office PC with a large monitor and with a single column on an iPhone. Bootstrap also adjusts to the orientation of the device automatically.

 By way of contrast, jQuery Mobile implements a progressive enhancement approach to Responsive design. Applications using this approach are built with a specific screen size.

We will cover Bootstrap in depth in *Chapter 5, Providing Cross-device Support with Responsive Design.*

The Geolocation API

One of the most fundamental requirements of any mobile ArcGIS app is its ability to enable users to find and show their current location on a map. This is where the Geolocation API comes in handy.

 The Geolocation API is a browser API that's part of JavaScript, but it is not related to ArcGIS.

On mobile devices, current location is derived from either GPS or cell tower triangulation. The Geolocation API taps into this data. Built-in security means that users need to provide permission before this functionality can be used in an application. Usually a popup will be displayed that will ask "The website *x* would like to use your current location". Clicking on the `Allow` button will give permission to the application. Building apps with the Geolocation API allows users to see and track their position on a map, and it is shown in the following screenshot:

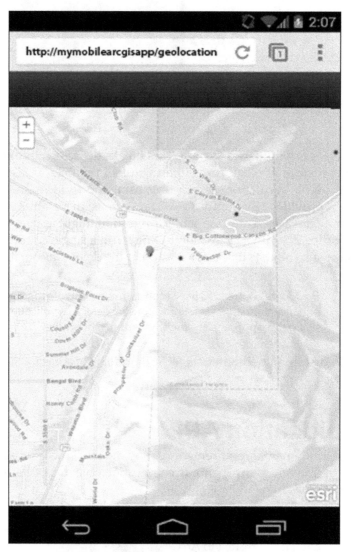

Showing current location using the Geolocation API

Most modern mobile browsers support the Geolocation API.

 Caniuse.com is an excellent resource that offers a quick way to see which browsers support the Geolocation functionality. You can find the information at http://caniuse.com/#search=geolocation.

The following code snippet shows a simple use of the API:

```
<p id="geolocate">Click the button to get your coordinates:</p>
<button onclick="getLocation()">Find my Location</button>
<script>
var x = document.getElementById("geolocate");
function getLocation() {
    if (navigator.geolocation) {
        navigator.geolocation.getCurrentPosition(showPosition);
    } else {
        x.innerHTML = "Geolocation is not supported by this browser.";
    }
}
function showPosition(position) {
    x.innerHTML="Latitude: " + position.coords.latitude +
    "<br>Longitude: " + position.coords.longitude;
}
</script>
```

This code snippet does 4 things:

- It checks to see whether Geolocation API is supported.
- If it is supported, it runs the getCurrentPosition() method. If it is not supported, it displays a message to the user.
- If the getCurrentPosition() method is successful, it returns a coordinates object to the function specified in the parameter, showPosition.
- The showPosition() function displays the latitude and longitude.

Summary

In this chapter, we reviewed the core building blocks of any mobile ArcGIS web application. The ArcGIS API is built using the Dojo framework. We provided you with a quick overview of Dojo, including its modules, widgets, and plugins. You will see these again when we start coding. Though jQuery Mobile is not a part of our focus in this book, it was worth mentioning since it is a good alternative to Dojo. The Geolocation API and the Bootstrap framework that were introduced in this chapter have set the stage nicely for later discussion.

This completes the two important foundation chapters. Now it's time to move to the exciting stuff. Time to start coding. In the next chapter, we will introduce code using many of the elements discussed in this chapter.

3

Building Your First Mobile ArcGIS Application

We've covered mobile development and how it is different from traditional web development. We also looked at some of the key JavaScript frameworks and toolkits. It's now time to build our first mobile ArcGIS application.

In this chapter, we will cover the following topics:

- Development and coding review
- The ArcGIS API for JavaScript API
- ArcGIS map layers
- Listening for map events
- JavaScript Geolocation API

Development and coding review

Before we move to the meat of this chapter, it is worth quickly reviewing some basic web development concepts and approaches. Specifically, let's consider the following:

- JavaScript Development Tools
- WebKit browsers
- HTML5, CSS3, and the viewport
- Web server setup

JavaScript development tools

Often, when developers embark on a new programming project, they turn to their favorite integrated development environment or IDE. There are many IDEs in the JavaScript world; these include intelliJ IDEA, Aptana, WebStorm, and Visual Studio. One of the beauties of JavaScript development is that any simple text editor will suffice. Notepad++ is a free and excellent source code editor for Windows. Chrome's DevTools is also very popular with JavaScript developers. This is a set of web authoring and debugging tools that is built into Google Chrome. To access Chrome's DevTools, download the current version of the Chrome browser.

 Chrome can be downloaded from `https://www.google.com/chrome/`.

As shown in following screenshot, clicking on the menu item in the top-right corner of the browser and navigating to **More Tools | Developer tools** will open DevTools (Shortcuts: *Ctrl + Shift + I* or *F12*):

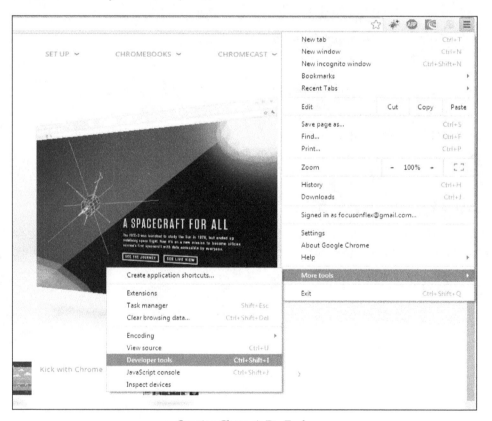

Opening Chrome's DevTools

Chrome's DevTools provides many debugging options that include breakpoints and object inspection. A new super useful feature is the device mode and mobile emulator. Using this option, you can test your mobile application on different simulated devices. Another very useful part of Chrome DevTools is remote debugging. This allows developers to debug live content on an Android device from any development machine.

 Find out more about remote debugging on Android using Chrome at `https://developer.chrome.com/devtools/docs/remote-debugging`.

The following screenshot illustrates debugging options in Chrome's DevTools to Mobile Emulator:

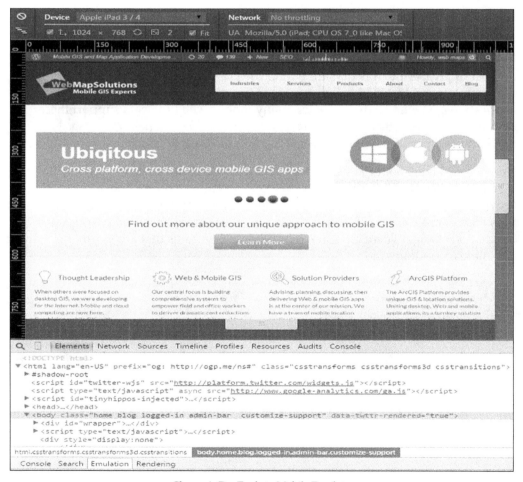

Chrome's DevTools to Mobile Emulator

It is worth getting to know Chrome's DevTools better. There are a plethora of tools provided to help mobile web developers. You can find out more at `https://developer.chrome.com/devtools`.

Next, let's discuss in brief the building blocks of web application development—HTML5, CSS3, and the viewport.

 For a deep dive into web ArcGIS development, see Eric Pimpler's book *Building Web and Mobile ArcGIS Server Applications with JavaScript*, available at `https://www.packtpub.com/hardware-and-creative/building-web-and-mobile-arcgis-server-applications-javascript`.

The base of all web applications is HTML. The new HTML standard is **HTML5**. This new specification defines how the tags and angle brackets, which make up HTML, interact with JavaScript through the **Document Object Model (DOM)**. There are a number of JavaScript APIs that are part of the HTML5 standard and these include the Geolocation API. **CSS3** is the new cascading stylesheet standard for styling web applications. As we will discuss later, many mobile web browsers support these new standards. Together, HTML5, JavaScript, and CSS3 will help to increase the flexibility and performance of your mobile mapping applications.

Now we will talk about the **viewport**. A browser identifies a page using the viewport: this is the viewable area of the canvas. The viewport controls how much a browser zooms into a page. On a mobile device, the viewport can be larger or smaller than the viewable area of the screen. This can cause problems with map layout and orientation change. There are some recommended settings that are worth mentioning and these are as follows:

```
<head>
<meta name="viewport"
  content="width=device-width,
  initial-scale=1.0,
  maximum-scale=1.0,
  user-scalable=no">
</head>
```

The width attribute sets the width of the canvas to the width of the device's screen. The initial-scale property controls the zoom level when the page is first loaded. By default the initial-scale is 1.0. A scale of 1.0 means that no zoom is applied to the view. The maximum-scale property controls how the users are allowed to zoom in or out of the page. Viewport zooming can be prevented by setting the user-scalable property to no, as we have done in this case. This way only the map will zoom with a finger pinch, for example, and not the full screen.

Not all browsers are built equally. Before we move to the basics of web development, let's consider their differences and the ever more popular WebKit engine.

WebKit and browsers

One of the more challenging aspects of web development is cross-browser compatibility issues. What works and looks good in one browser might not look good in another. As we mentioned earlier, some JavaScript frameworks help but do not remove all of these challenges. Web standards are applied differently between browsers.

 Web standards are the formal, non-proprietary standards that define and describe aspects of the web. They have become a set of standardized best practices for building web sites.

WebKit is one of the most standards-compliant browser rendering engines that is currently available. A rendering engine is the engine that a browser uses to render or read and display the HTML in a web page. Webkit fully supports the new HTML and CSS standards: HTML5, JavaScript, and CSS3. The good news is that most mobile browsers use WebKit as their rendering engine. One of the exceptions is Windows Internet Explorer, which uses the Trident engine. The large number of standards-compliant browsers means that cross-browser compatibility issues are less important now than they were in the past, particularly to mobile web developers. However, they still cannot be ignored.

 Unfortunately, the Android OS continues to be problematic for debugging because the default browser used isn't consistent. You can find more information about the Android OS at http://slides.com/html5test/the-android-browser#/.

Before we step into coding, let's cover one other area: mirroring the environment in which your mobile application will ultimately reside.

Web server setup

We've spoken about IDEs and development environments. Although it is your choice, we suggest that you install a web server locally. This way you can develop and test your application in a real environment. Apache is one of the most popular web servers. It's also free. Another good option for Windows machines is **Internet Information Server (IIS)**.

 You can download the Apache web server from `http://httpd.apache.org`. On Windows, at least, we have found that the MSI Installer package, which usually resides in the binaries directory, is the simplest way to install the server.

Once you have downloaded the Apache installer, double-clicking on the package will begin the installation process. You will be walked through a series of steps by the installer. Some steps will require you to make a choice. For a basic installation, the default is usually the best option.

 The default folder under which Apache is installed can be found by navigating to `c:|\\192.168.0.200\ProgramProgram Files|Apache Software Foundation|Apache2.2`. During the installation process, it is a good idea to change this default location to something simpler, such as `c:\webroot`.

We now have our development tools and environment set up. Let's build our first ArcGIS mobile web application.

The ArcGIS API for JavaScript API

The increasing popularity of JavaScript has encouraged Esri to assemble a large and experienced development team that is focused on the ArcGIS JavaScript API. They have built a feature-rich set of classes that are organized into modules.

 All the documentation relating to the ArcGIS JavaScript API can be found at `https://developers.arcgis.com/javascript/`. Here, you can find the samples, the API reference, tutorials, and much more.

For ArcGIS developers, Esri has released a new compact build. This is smaller than the standard build, which is perfect for both mobile developers and those who wish to use JavaScript libraries other than Dojo.

Web Optimizer is also worth mentioning. This allows developers to customize their own build and can significantly reduce the size of the library even further. As discussed in *Chapter 7*, *Developing Hybrid ArcGIS Mobile Applications with PhoneGap* this is also important for Hybrid apps because it provides a single file that can be hosted locally on the device. For more information, go to `http://jso.arcgis.com/`.

As you will see in the code examples, reference to the compact build includes the compact wording, for instance, `http://js.arcgis.com/3.11compact`.

The compact build is smaller than the standard build for two main reasons:

1. Widgets, or dojo digits, are not automatically loaded. A number of widgets are needed in a standard web application, but they are unnecessary in a mobile application.

2. Only 32 modules are loaded in place of the standard 80 modules. Mobile developers will need to use `require()` to load any of the additional modules.

The recommended approach to access the API is to include the following link and script tag in your page:

```
<link rel="stylesheet" href="http://js.arcgis.com/3.11/esri/css/esri.
css">
<script src="http://js.arcgis.com/3.11compact/"></script>
```

This approach delivers the libraries via a **content delivery network (CDN)**. CDNs are collections of geographically distributed web servers that deliver content efficiently. In the preceding code sample, we loaded the compact build and the `esri.css`. This CSS file contains the CSS for various widgets, including the CSS for the map.

The ArcGIS JavaScript API provides support for many languages. To use one of these, simply run your application with the appropriate locale. You can read more about this at `https://developers.arcgis.com/javascript/jshelp/localization.html`.

Now it's time to write some code. Follow these steps to get started:

1. Open your favorite IDE or Notepad++.

2. Now let's start with the basic HTML page layout. You will notice reference to the viewport, stylesheet, and the compact build:

```
<!DOCTYPE html>
<html>
  <head>
    <meta http-equiv="Content-Type" content="text/html;
charset=utf-8">
```

```
    <meta name="viewport" content="initial-scale=1, maximum-
scale=1,user-scalable=no"/>
    <title>Simple Map</title>
    <link rel="stylesheet" href="http://js.arcgis.com/3.11/esri/
css/esri.css">
    <script src="http://js.arcgis.com/3.11compact/"></script>
  </head>
  <body>
    <div id="map"></div>
  </body>
</html>
```

Notice that we have included a <div> tag in this code. This will hold the map that we are about to create.

1. Next, let's load the Dojo ArcGIS Mapping module using require():

```
<script>
  require(["esri/map", "dojo/domReady!"],
    function(Map){
    });
  });
</script>
```

2. This gives us our basic structure. Now let's create a new Map object and set the center of the map, zoom level, and basemap:

```
<script>
require(["esri/map", "dojo/domReady!"], function(Map) {
  var map = new Map("map", {
    center: [-118, 34.5],
    zoom: 7,
    basemap: "streets"
  });
});
</script>
```

 The dojo/domReady! plugin ensures that all the HTML elements have been loaded.

3. That's it. The completed code will look like the following:

```
<!DOCTYPE html>
<html>
  <head>
    <meta http-equiv="Content-Type" content="text/html;
charset=utf-8">
```

```
    <meta name="viewport" content="initial-scale=1, maximum-
scale=1,user-scalable=no"/>
    <title>Simple Map</title>
    <link rel="stylesheet" href="http://js.arcgis.com/3.11/esri/
css/esri.css">
    <script src="http://js.arcgis.com/3.11compact/"></script>
    <script>
    require(["esri/map", "dojo/domReady!"], function(Map) {
        var map = new Map("map", {
center: [-118, 34.5],
zoom: 7,
basemap: "streets"
        });
    });
    </script>
</head>
<body>
    <div id="map"></div>
</body>
</html>
```

Save this code as `firstMobileMap.html` and copy it to `c:\webroot\htdocs`. Now open a web browser and type `http://localhost/firstMobileMap.html`. Eureka! we can view, pan, and zoom our first map, which is shown below:

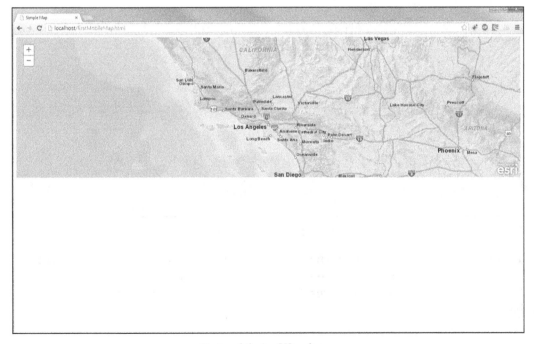

First mobile ArcGIS web map

However, we seem to have a problem. The map only fills the top half of the screen. It's time to add some styling. Using the loaded `esri.css` style sheet, we can style the application so that the map fills the screen. We need to add the following piece of code to our code sample directly below our `esri.css` style sheet reference:

```
<style>
  html, body, #map {
    height: 100%;
    width: 100%;
    margin: 0;
    padding: 0;
  }
</style>
```

The following screenshot illustrates the styled mobile ArcGIS web map:

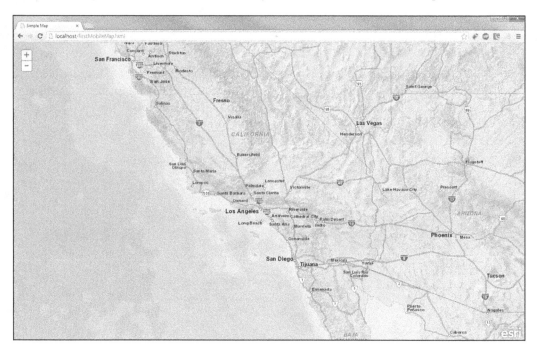

First styled mobile ArcGIS web map

There! Now that's a little better. The esri.css is very useful for styling your ArcGIS mobile applications. It is a sizeable file, but you will often see reference to it in Esri JavaScript guides and code samples. In our simple case, we have set the width and height of the map to 100%.

ArcGIS map layers

We have made a good start. We have our first displayed basemap and it's time to talk about map layers.

Map layers provide context and meaning to any GIS mobile application. They also set the stage for further analysis. Maps are made up of different types of data and this data can be served up through different spatial web services. For ArcGIS, map layers are actually RESTful web services. The web service for the marathon layer that we will be using in this chapter can be seen in the following screenshot:

ArcGIS REST Services Directory

Home > services > Boston_Marathon (FeatureServer) > Boston_Marathon

JSON

Layer: Boston_Marathon (ID:0)

View In: ArcGIS.com Map

Name: Boston_Marathon

Display Field:

Type: Feature Layer

Geometry Type: esriGeometryPolygon

Description:

Copyright Text:

Min. Scale: 0

Max. Scale: 0

Default Visibility: true

Max Record Count: 2000

Supported query Formats: JSON

Extent:

    XMin: -19839092.3000987
    YMin: 2145729.68522357
    XMax: -7454985.15052452
    YMax: 11542624.9015675
    Spatial Reference: 102100

ArcGIS feature layer REST interface

There are three primary layer types: dynamic map service layers, Tile Map Service layers, and feature map service layers. A dynamic map service layer creates vector data dynamically. As the user navigates the map, the layer requests new map images to be displayed. An ArcGIS Tiled Map Service layer displays the map content from an ArcGIS Server map service that has been cached (tiled). A cached map service contains pre-generated map tiles. Our marathon layer is a feature map service. This is raw layer data that is passed for rendering to the browser on the mobile device. Clients can execute queries against feature map services to get features and perform edits that can be applied to the server. This is an enhanced type of graphics layer that can be used to display points, lines, or polygons from a map service or a dedicated feature layer. If a feature layer is enabled for editing, you can also add, update, or delete data.

As mentioned, Feature layers can be of different types: point, line, or polygon. Let's extend our code and overlay a feature layer on our basemap. Esri has made a number of published feature layers publicly available and we will use their Boston Marathon polygon layer. This shows the percentage of runners who finished by US State. The feature layer JavaScript syntax is as follows:

```
var marathon = new FeatureLayer("http://services.arcgis.
com/V6ZHFr6zdgNZuVG0/arcgis/rest/services/Boston_Marathon/
FeatureServer/0", {
    mode: FeatureLayer.MODE_ONDEMAND,
    outFields:["*"]
});
```

By breaking this down, we need to create a `FeatureLayer` object. The `FeatureLayer` constructor takes the layer URL, the mode, and outfields or attributes to display. In this case, we wish to view all the attributes.

> A feature layer draws quickly in on-demand mode because the features are retrieved from the server as needed. There are other mode options including snapshot that retrieves all the features from the associated layer resource.

Once we have created the `FeatureLayer` object, we need to add this to the map. This is done using the map object's `addLayers` method:

```
map.addLayers([marathon]);
```

Note that the `addLayers` method is for multiple layers and it takes an array denoted by `[]` as a parameter. A quick addition is needed to our `require()` function to include the `FeatureLayer` module:

```
<script src="http://js.arcgis.com/3.11compact/"></script>
<script>
require(["esri/map", "esri/layers/FeatureLayer", "dojo/
domReady!"], function(Map, FeatureLayer) {
    var map = new Map("map", {
center: [-118, 34.5],
zoom: 7,
basemap: "streets"
    });
var marathon = new FeatureLayer("http://services.arcgis.
com/V6ZHFr6zdgNZuVG0/arcgis/rest/services/Boston_Marathon/
FeatureServer/0",
    {
mode: FeatureLayer.MODE_ONDEMAND,
outFields:["*"]
    });
map.addLayers([marathon]);
    });
    </script>
```

Now this is perfect. Let's test the code. Update the file, `firstMobileMap.html`, with the new additions and you should see the Marathon polygon layer overlaid on the streets basemap.

Note that an excellent resource provided by Esri for writing and testing your code is the JavaScript Sandbox. This can be found at `http://developers.arcgis.com/en/javascript/sandbox/sandbox.html`.

Listening for map events

We have now set up our basic map structure. Our application now loads a basemap and overlays a feature layer. Often users want to select a feature in a layer and view its attributes. This is where listening for and capturing events comes in handy. Let's add a listener to the marathon feature layer as follows:

```
marathon.on("click", myClickHandler);
```

This is a click event so that anytime a user clicks or taps the layer we want to pass the event to a function called myClickhandler(). An event object is passed to the event handler as this function is called. What we would like to do is view the attributes of the feature object—in this case, the state that the user selected.

```
<script>
    require(["esri/map", "esri/layers/FeatureLayer","dojo/on",
"dojo/domReady!"], function(Map, FeatureLayer, on) {
        var map = new Map("map", {
    center: [-118, 34.5],
    zoom: 7,
    basemap: "streets"
        });
    var marathon = new FeatureLayer("http://services.arcgis.
com/V6ZHFr6zdgNZuVG0/arcgis/rest/services/Boston_Marathon/
FeatureServer/0",
{
    mode: FeatureLayer.MODE_ONDEMAND,
        outFields:["*"]
});
marathon.on("click", myClickHandler);
map.addLayers([marathon]);
//Marathon layer click handler
function myClickHandler(event)
{
alert(JSON.stringify(event.graphic.attributes));
};
});
    </script>
```

 The myClickHandler function is passed as a click event object. Load the web page and open Chrome's DevTools. Put a breakpoint at myClickHandler and then add a watch expression called event. You should then be able to see the contents of the click event object. The event. graphic.attributes lists the attributes of the selected graphic. Later in this chapter, you can see the DevTools screenshot that shows object introspection.

Notice that we need to convert the attributes object to a string. We use JSON. stringify() to do this conversion. In the following screenshot, you can see how the web application now looks on a Nexus 4 using the DevTools emulator:

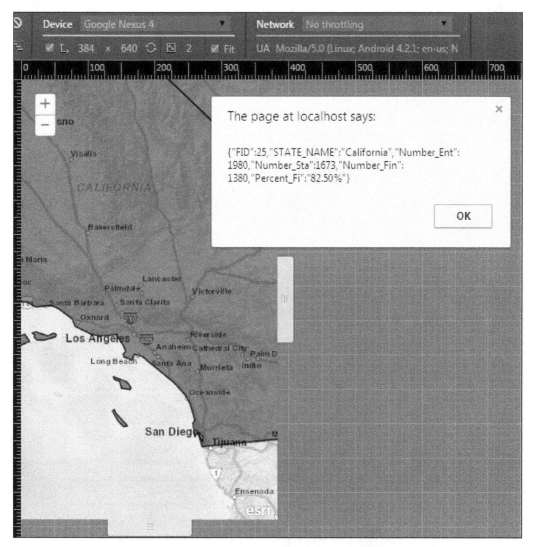

An event handler for layer tap showing feature attributes

For more information, read the Chrome developer documents at
`https://developer.chrome.com/devtools/docs/javascript-debugging`.

One incredibly useful tool is object introspection. If we wish to see the contents of the event object passed to our `myClickHandler` function, we can set a breakpoint inside the function and make the event object our watch expression.

In the following screenshot, we have shown you how this can be done using DevTools in Chrome:

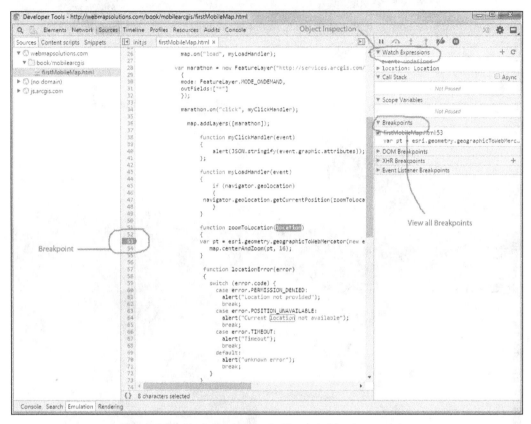

Chrome's DevTool showing breakpoint and object introspection

We've only touched on event listeners in this chapter. There are many layers in the ArcGIS JavaScript API that include load, pan, zoom, addLayer and basemap-change. These will come in handy when you write more mobile ArcGIS applications and need to detect both map changes and user actions.

JavaScript Geolocation API

It's time to add a very mobile-specific element to our code — geolocation or "zoom to my current location" functionality. We discussed the Geolocation API in *Chapter 2, Understanding Mobile Frameworks and APIs*. Let's now put it to use. First, we will need to set an event listener, in this case, a map load listener:

```
map.on("load", myLoadHandler);
```

Inside the `myLoadHandler` function we will reference the Geolocation API. We will detect whether or not the `navigator.geolocation` object is available. This is a best practice because not all browsers support geolocation-based navigation. The position object will return `coords.latitude`, `coords.longitude`, and `coords.accuracy`. It might also return `coords.altitude`, `coords.altitudeAccuracy`, `coords.heading`, `coords.speed`, and timestamp. Note that this varies by browser. We also need to handle any errors if the object is not available with the `locationError()` function. The following code snippet determines geolocation-based navigation object availability:

```
function myLoadHandler(event)
{
  if (navigator.geolocation)
  {
 navigator.geolocation.getCurrentPosition(zoomToLocation,
locationError);
  }
}
```

In the `myLoadHandler` function, we first check whether the navigator.geolocation object is available and, if it is, then we use the coordinates contained there to generate a point object.

If the accuracy of the mobile GPS is a concern, it can be resolved by the following line of code:

```
if (location.coords.accuracy <= 500){}
```

> The value provided here, 500 in our case, will depend on your requirements. This can be particularly useful if your application needs to geolocate only if high accuracy is a requirement. As a best practice, if accuracy is a requirement, the `enableHighAccuracy` property should be set to true. For more information, go to `http://dev.w3.org/geo/api/spec-source.html#high-accuracy`.

Note that the ArcGIS JavaScript API provides a very useful coordinate conversion function. The `geographicToWebMercator` converts the geographic coordinate data returned by the Geolocation API to Web Mercator, which is the default projection of ArcGIS maps. This requires one to import the `esri/geometry/webMercatorUtils` modules.

> For more information on map projections, see the article at `http://support.esri.com/ja/knowledgebase/techarticles/detail/23025`.

As shown in the `zoomToLocation` function of the following code, we use point object to center the map using the `map.centerAndZoom()` function. In the sample code, the zoom level is set to 16, so the map zooms in automatically to the user's location:

```
function zoomToLocation(location) {
    var pt = esri.geometry.geographicToWebMercator(new esri.geometry.
Point(location.coords.longitude, location.coords.latitude));
    map.centerAndZoom(pt, 16);
}
```

Our error handler will provide the user with specific information on why geolocation failed. To do this, use the following code snippet::

```
function locationError(error) {
    switch (error.code) {
        case error.PERMISSION_DENIED:
            alert("Location not provided");
            break;
        case error.POSITION_UNAVAILABLE:
            alert("Current location not available");
            break;
        case error.TIMEOUT:
            alert("Timeout");
            break;
        default:
            alert("unknown error");
            break;
    }
}
```

If you wish to track the changing position of a user, you can use the `watchPosition()` function:

```
navigator.geolocation.
watchPosition(showLocation, locationError);
```

This is useful for waypoints, as an example.

Let's pull this all together into our file, `firstMobileMap.html`. We have a copy of the final code at `http://webmapsolutions.com/book/mobilearcgis/chapter3/firstMobileMap.html`.

Test the code. You will find that the map loads with its center at your current location. Note that you will be testing on your home computer and not a mobile device; this should still zoom to your current location without a problem.

Summary

In this chapter, you learned the basics of developing an ArcGIS mobile web application. Before coding, it is often a good idea to set up your environment properly. This was why we spent time considering IDEs, debugging tools, and mobile device emulators. Since ultimately your users will be accessing your mobile application via a web server, we think that having Apache or an equivalent software installed locally would be a good idea for ongoing development testing. The core of the chapter was devoted to coding, walking through, and step-by-step building of your first mobile application using the ArcGIS API. We began by putting in place the core code elements, which allowed us to load a basemap. From there, we added a feature layer, event handler, and finally a geolocation functionality. In the next chapter, we will develop this code further and add some more advanced functionality.

4
Advancing the Basic Mobile ArcGIS Application

In the previous chapter, we built our first mobile ArcGIS app. The application loaded a basemap and a feature layer, and we added geolocation functionality. Next, we will jump into some more advanced coding. This is where the fun begins. We will build ArcGIS apps that include some popular mobile tools.

In this chapter, we will cover the following topics:

- Adding popular tools
- Feature popups
- Adding a legend
- Finding features
- Address Search

Getting started

In *Chapter 3, Building Your First Mobile ArcGIS Application*, we added a layer of data provided by an ArcGIS feature service. This is a special type of service where the raw layer data is passed to the mobile browser for rendering. This is different to the other commonly used ArcGIS map services — dynamic and tiled — that provide map images. You can interact directly with features generated by an ArcGIS feature service programmatically. Using feature layers, and setting the correct "Mode", can avoid the need for a round-trip request-response to ArcGIS when we pan, zoom, or wish to see a features attribute.

The following screenshot illustrates the various types of layers in ArcGIS:

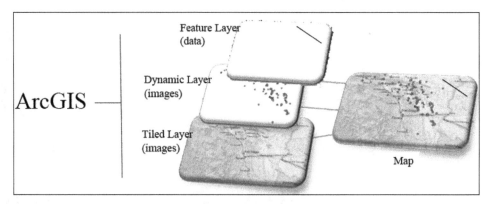

ArcGIS layer types

Avoiding server requests has potential advantages, particularly relating to performance. In this chapter, we will remain focused on feature layers most of the the code examples take advantage of in-memory feature layer data.

Adding popular tools

In this chapter, we will advance our ArcGIS API knowledge and skills by building some popular mobile tools. Each of the code examples will be built using a core code base.

Esri provides hundreds of excellent ArcGIS JavaScript samples that are a great source for all ArcGIS web developers. You can access the full set of samples at https://developers.arcgis.com/ javascript/jssamples/. Our starter code sample is available at http://www.webmapsolutions.com/book/mobilearcgis/ chapter4/MobileMapBase2.html.

Feature popups

Once mobile GIS users can see their data on a map, they often wish to discover more about the displayed data. The layers that are overlaid on a basemap contain geographic features. These are usually represented by one of three forms: points, lines, or polygons:

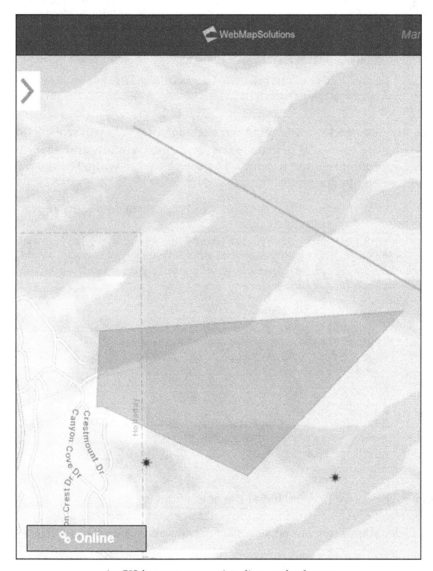

ArcGIS feature types – points, lines, and polygons

Using a popup window, Esri makes it simple to display details when a user taps a feature. A special `PopupMobile class` has been introduced for mobile development.

When publishing layers in ArcGIS to use in a mobile application, make sure that the size the size of the symbol associated with each feature is big enough to work for a finger tap. Both points and lines can be difficult to tap if the size of a point or the width of a line feature is too small. It is good to experiment here, but often point features are best with a diameter of over 18 pixels, and lines with a width of over 3 pixels. If you are consuming a feature service and the symbols aren't large enough to tap on, it is possible to override the feature service symbology with a Renderer. More on Renderers can be found at the following link:

```
https://developers.arcgis.com/javascript/jshelp/
inside_renderers.html
```

This is the first use of a Dojo widget in our code samples. We will work with two different widgets: `PopupMobile` that creates the popup window and `PopupTemplate` that defines the contents of the window. The `PopupMobile` widget has two style sheets that we need to define:

```
<link rel="stylesheet" type='text/css' href='http://js.arcgis.
com/3.11/esri/dijit/css/Popup.css'/>
<link rel="stylesheet" type='text/css' href='http://js.arcgis.
com/3.11/esri/dijit/css/PopupMobile.css'/>
To include these Esri widgets and supporting modules, we need to
update our require function:
require([

        "esri/map", "esri/layers/FeatureLayer", "esri/symbols/
SimpleFillSymbol", "esri/symbols/SimpleLineSymbol",
        "esri/dijit/PopupMobile", "esri/dijit/PopupTemplate", "esri/
Color",
        "dojo/on", "dojo/domReady!"],
        function(
        Map, FeatureLayer, SimpleFillSymbol, SimpleLineSymbol,
        PopupMobile, PopupTemplate, Color,
        on
        ) {
```

You will notice that we have included the new widgets `PopupMobile` and `PopupTemplate`, plus `SimpleFillSymbol`, `SimpleLineSymbol`, and `Color`. As you will see later, these are needed to construct our `PopupMobile` object.

Since our marathon feature layer is of the polygon type, it would be nice when a user taps a feature if we highlight both the border and body of the polygon. We will create the `PopupMobile` widget with the selected feature highlighted as follows:

```
var popup = new esri.dijit.PopupMobile({fillSymbol: new esri.symbol.
SimpleFillSymbol(esri.symbol.SimpleFillSymbol.STYLE_SOLID, new esri.
symbol.SimpleLineSymbol(esri.symbol.SimpleLineSymbol.STYLE_SOLID, new
esri.Color([255,0,0]), 2), new esri.Color([255,255,0,0.25]))}, dojo.
create("div"));
```

The border is set using `SimpleLineSymbol`, to a solid red line (255, 0, 0) and, with `SimpleFillSymbol`, the polygon body to a solid yellow (255, 255, 0). You will see variations on this code later in this chapter as we highlight different features.

Next, the Map's `infoWindow` constructor option needs to be set to reference our new `PopupMobile` widget:

```
//map object
   map = new Map("map", {center: [-118, 34.5], zoom: 7,
basemap: "streets", infoWindow: popup});
```

As we mentioned earlier, `PopupTemplate` sets the content of the `PopupMobile` window. `PopupTemplate` allows you to quickly specify content that includes field information, text description, images, and charts.

> A feature's fields are its attributes. For example, a parks polygon layer might have features whose fields include facilities (playgrounds, soccer fields, fishing), the park size, pavilion booking information, and so on.

In this example, for simplicity, we will display a title, the state's name, and the percentage of racers who finished the race:

```
var template = new esri.dijit.PopupTemplate({
   title: "Boston Marathon 2013",
   description: "{STATE_NAME}: {Percent_Fi} of starters finished"
   });
```

Note the syntax; we referenced the REST endpoint field names within curly brackets, for example, {STATE_NAME}.

It is often useful to look at the REST endpoint for published ArcGIS layers. Copy and paste the following marathon polygon layer link that we are using in our code examples into your browser window:

```
http://services.arcgis.com/V6ZHFr6zdgNZuVG0/arcgis/
rest/services/Boston_Marathon/FeatureServer/0
```

Here you will find all the information about this (ArcGIS REST) service. This includes the layer type, drawing information, field names, and much more:

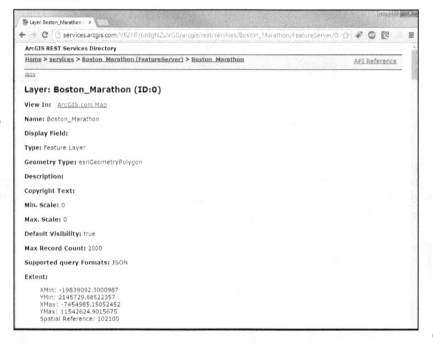

The following is the screenshot from the REST directory showing the field names:

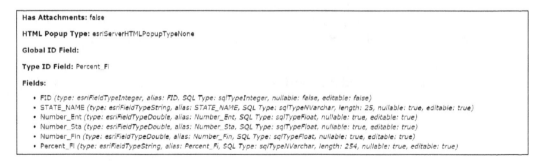

ArcGIS feature layer REST endpoint

Finally, we need to set the `infoTemplate` constructor option for our marathon feature layer to our new `PopupTemplate`:

```
var marathon = new FeatureLayer("http://services.arcgis.
com/V6ZHFr6zdgNZuVG0/arcgis/rest/services/Boston_Marathon/
FeatureServer/0",
{
mode: FeatureLayer.MODE_ONDEMAND,
outFields:["*"],
infoTemplate: template
});
```

You can view this example at:

http://www.webmapsolutions.com/book/mobilearcgis/chapter4/
MobileMapPopUp.html.

Notice how the selected polygon stands out and, by using `PopupMobile`, we get a shortened popup that opens a new screen with details when users tap the arrow button:

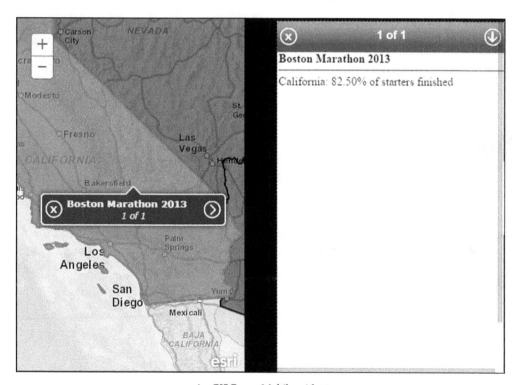

ArcGIS PopupMobile widget

We can go one step further with this example. `PopupTemplate` is extremely flexible. If we want to add charting to `MobilePopUp,` that can be done. By extending the code base, it is possible to add bar charts. This can be seen in the following screenshot:

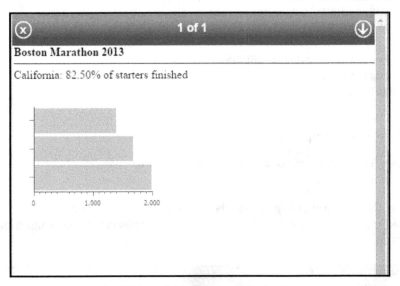

ArcGIS PopupMobile widget with charting

The extended version of this example is available at `http://www.webmapsolutions.com/book/mobilearcgis/chapter4/MobilePopUpChart.html`.

Adding a legend

A user can tap a feature and see the details in a mobile-enabled popup. It would be nice to differentiate more easily between the respective map and feature layers. To do this, a legend would be an excellent addition. For the next code sample, we will again use our starter code sample: `http://www.webmapsolutions.com/book/mobilearcgis/chapter4/MobileMapBase2.html`. Let's begin by showing you a screenshot of the application that you are about to build:

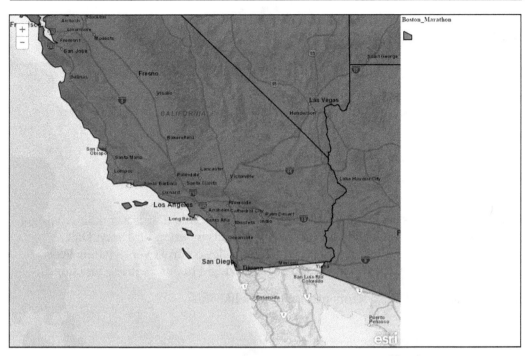

ArcGIS Legend widget

Notice the layout. The screen is subdivided into two: a map and a legend panel. First, we need to update our code to include this new layout. In this code example, we need to update the style first by changing the map width and adding `rightPane`:

```
<style>
    html, body, #map {
  height: 100%;
  width: 98%;
  margin: 0;
  padding: 0;
    }
    #rightPane {
  width: 20%;
height: 100%;
  background: White;
  position: absolute;
  top:1px;
  right:1px;
  z-index: 99;
  overflow-x: hidden;
  overflow-y: auto;
    }
</style>
```

You need to add two new library elements to the `require` section; namely, `BorderContainer` and `ContentPane`:

```
require(["esri/map", "esri/layers/FeatureLayer",
        "dijit/layout/BorderContainer", "dijit/layout/ContentPane",
"esri/dijit/Legend",
        "dojo/on", "dojo/domReady!"],
        function(
        Map, FeatureLayer,
        BorderContainer, ContentPane, Legend,
        on
        ) {
```

These are containers for the map and legend. `BorderContainer` is a container widget which provides an easy way to lay out page elements in a standard way. Dijit `ContentPane` is a basic layout widget which provides control over dynamic Web content. In addition we have added the Legend widget to our `require` function.

The body of our HTML document should look like this:

```
<body>
  <!-- body content -->
  <div id="content"
      data-dojo-type="dijit/layout/BorderContainer"
      data-dojo-props="design:'headline', gutters:true"
    style="width: 100%; height: 100%; margin: 0;">
  <!-- Right side legend panel -->
      <div id="rightPane"
          data-dojo-type="dijit/layout/ContentPane"
        data-dojo-props="region:'right'">
              <div data-dojo-type="dijit/layout/ContentPane"
              data-dojo-props="title:'Pane 2'">
          <div id="legendDiv"></div>
        </div>
      </div>
      <!-- map -->
      <div id="map"
          data-dojo-type="dijit/layout/ContentPane"
          data-dojo-props="region:'center'"
        style="overflow:hidden;">
    </div>
  </div>
  </body>
```

Here, we set the div tag to a `data-dojo-type` and the required container. You can see this first phase of the legend coding at:

```
http://www.webmapsolutions.com/book/mobilearcgis/chapter4/
MobileMapLegend1.html.
```

Let's now add a legend widget to the right pane. Inside the `init()` function, we need to add the following piece of code:

```
//When layer loaded add legend to right pane
    map.on("layers-add-result", function (evt) {
            var layerInfo = dojo.map(evt.layers, function (layer,
index) {
                return {layer:layer.layer, title:layer.layer.name};
            });
            if (layerInfo.length > 0) {
              var legendDijit = new esri.dijit.Legend({
                map: map,
                layerInfos: layerInfo
              }, "legendDiv");
              legendDijit.startup();
            }
    });
```

When each overlay layer loads (we have only one feature layer in our example), populate `layerinfo` with the map layer and layer name.

 Though a feature layer is published based on type (point, line, or polygon), a specific type can be symbolized in different ways. For example, a point layer can use different symbols to differentiate points based on a particular attribute.

Then, create a new legend widget passing in `layerinfo`, the map object, and the display `div` tag. We finalize the creation of the legend widget using `legendDijit.startup();`.

There we have it! The right-hand side pane should show the marathon polygon layer legend.

 All the feature layers that have been included in this map, using this code, are listed in this right-hand side pane.

This second phase of legend coding can be seen at:

```
http://www.webmapsolutions.com/book/mobilearcgis/chapter4/
MobileMapLegend2.html.
```

Finding features

So far, we have provided tools that help users to learn more about a layer and its features. Next, let's consider how we can find features. You can search a feature layer locally in memory or make a REST query to ArcGIS Online or ArcGIS Server. In this case we will make a remote call. We will again build from our base code. We have a number of goals for this feature search tool. We'd like a user to be able to type in a feature by name and automatically make the map zoom/pan to the highlighted feature.

It is often a good idea to run some queries manually against the feature layer of interest. In the case of our marathon layer, you can test out feature queries at:

```
http://services.arcgis.com/V6ZHFr6zdgNZuVG0/arcgis/
rest/services/Boston_Marathon/FeatureServer/0/query.
```

It is possible to return queries using GeoJSON. For more on feature service queries see the following link:

```
http://resources.arcgis.com/en/help/arcgis-
rest-api/index.html#/Query_Feature_Service_
Layer/02r3000000r1000000/.
```
In the preceding link, you can test the feature queries with sample parameters. For example, where 1=1, and outfields=*.

Again, beginning with our starter code sample, we need to update our require section:

```
require([
        "esri/map", "esri/layers/FeatureLayer", "esri/symbols/
SimpleFillSymbol", "esri/symbols/SimpleLineSymbol",
        "esri/tasks/query", "esri/geometry/Polygon", "esri/graphic",
        "esri/dijit/PopupMobile", "esri/dijit/PopupTemplate", "esri/
Color",
        "dijit/layout/BorderContainer", "dijit/layout/ContentPane",
        "dojo/on", "dojo/domReady!"],
        function(
        Map, FeatureLayer, SimpleFillSymbol, SimpleLineSymbol,
        Query, Polygon, graphic,
        PopupMobile, PopupTemplate, Color,
        BorderContainer, ContentPane,
        on
        ) {
```

We've added the query module and the modules that are needed to highlight the feature. We have also included the container widgets that were used in the legend example. In this case, instead of a right-hand side pane, we will add a top pane. This holds the feature input text box and the find button. Similar to other samples, a link to this code sample is provided at the end of this section and it can be reviewed there.

 To see the source code of any web page, simply right-click and, in the popup menu, select view page source in Chrome or an equivalent browser:

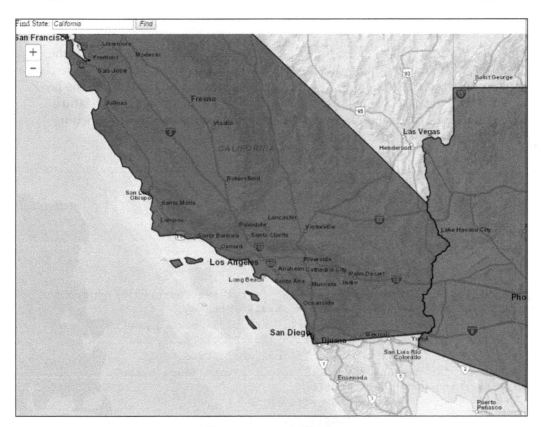

View page source in Chrome

Next, we need to add two new functions. The first executes the find feature query after a user has tapped on the Find button shown in the preceding screenshot. The code snippet for this is as follows:

```
//Execute find feature query
function findFeature(searchText) {
  var query = new esri.tasks.Query();
  query.where = "STATE_NAME = '" +  searchText + "'";
  marathon.queryFeatures(query, function (featureSet) {
    fetchRecords(featureSet);
      });
}
```

The `findFeature(searchText)` function is passed as `searchText`, which is the user's search input. Our query object needs a where statement. This is a where clause for the query filter. Any legal SQL where clause operating on the fields in the layer is allowed. In our case, we will use the `STATE_NAME` attribute to find the feature based on name. We execute the query and pass the results to our next function, `fetchRecords()`, which is as follows:

```
//Display the results of the find feature query
function fetchRecords(featureSet) {
  //Clear the graphics layer
  map.graphics.clear();
  //If we get a result set
  if (featureSet.features.length > 0) {

//Highlight the feature
      var polygonJson  = {"rings":featureSet.features[0].geometry.
rings,"spatialReference":map.spatialReference};
      map.graphics.add(new esri.Graphic(new esri.geometry.
Polygon(polygonJson), new esri.symbol.SimpleFillSymbol(
          esri.symbol.SimpleLineSymbol.STYLE_SOLID,
          new esri.symbol.SimpleLineSymbol(esri.symbol.
SimpleLineSymbol.STYLE_SOLID,
            new esri.Color([0, 0, 0, 1]), 1),
          new esri.Color([255, 0, 0, 0.2]))
      ));
  }
  }
```

There are a number of parts to our `fetchRecords()` function:

1. We first clear the graphics layer to remove the results of any previous query.

2. Next, we check whether we have a result set from our new query, if we create an extent object based on the result set.

3. We then set the map to this new extent object and this effectively zooms/pans the map to our feature of interest.

4. The final section of this code snippet shows how we highlight the state by adding a graphic to the map. It resembles the earlier highlighting code that we used for popups. Polygon geometries in ArcGIS are defined by a single or multiple rings. This is an array of three or more points. Read more about polygon features in the ArcGIS API documents at `https://developers. arcgis.com/javascript/jsapi/polygon-amd.html`.

An example of the find feature is shown in the following image:

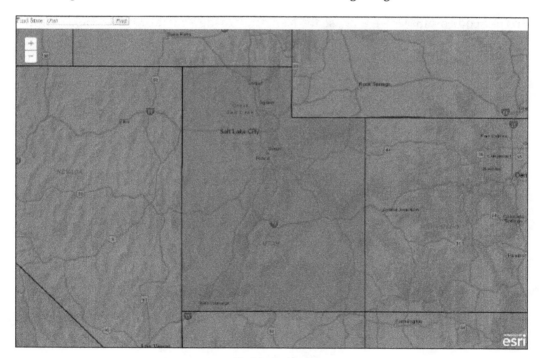

Find feature tool

We have also added a header with search input box to the page. See the application and code sample at:

`http://www.webmapsolutions.com/book/mobilearcgis/chapter4/ MobileMapFindFeature.html`.

Address search

Another commonly requested mobile tool is an address search. This takes us from directly working with a feature layer to using an external Geocoding service.

Esri's Geocoding Service can be used to convert an address or a place's name to latitude and longitude. For example, if a user knows an address and wants to put it on a map, geocoding can be used to find the coordinates. In our code example, we will use the free Geocode services that can be found at:

`http://tasks.arcgis.com/ArcGIS/rest/services/WorldLocator/GeocodeServer`. This is free only if you are not doing batch geocoding and are not saving the results.

Read more about Esri's Geocoding service at:

`https://geocode.arcgis.com/arcgis/index.html`.

For this code example, we will use the same basic code structure that we used for the find feature, in our require function, but we will replace "esri/tasks/query" with "esri/tasks/locator".

Though it is not discussed here, Esri's Geocoder module "esri/dijit/Geocoder" is also worth considering. You can read more about this at `https://developers.arcgis.com/javascript/jsapi/geocoder.html`.

The following screenshot shows the Find Address application we are about to build:

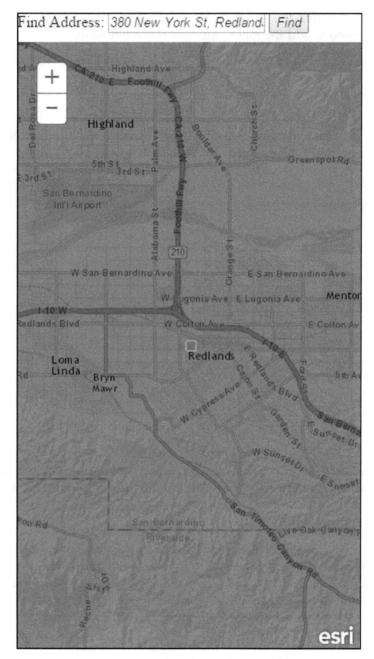

Find Address tool

We first need to initialize the geocoder and add a listener for geocoder responses. To do this, we need to add the following code to the init function:

```
locator = new Locator("http://geocode.arcgis.com/arcgis/rest/services/
World/GeocodeServer");
//add handler for button click
locator.on("address-to-locations-complete", showResults);
```

The showResults function handles the geocoder response. The "address-to-locations-complete" event can be used to find candidates for both single and multiple addresses. The findAddress(searchText) function, shown in the following code, is used to generate the geocoder request. This function is called when the user taps the **Find** button and the user address input is passed to this function as searchText:

```
//Execute find address request
function findAddress(searchText) {
//Clear any existing graphics
  map.graphics.clear();
    var address = {
    "SingleLine": searchText
    };
    locator.outSpatialReference = map.spatialReference;
    var options = {
    address: address,
    outFields: ["Loc_name"]
    };
        locator.addressToLocations(options);
}
```

We need to pass address and outFields data objects to our locator object. The address data object contains properties representing the various address fields accepted by the corresponding Geocode service, while outFields is a list of fields included in the returned resultset.

With the following code, we can illustrate the geocode response handler. Here, we first check whether we have a resultset. Then, we set the map extent to correspond to this resultset to automatically pan/zoom the map. Next, we add a graphic at the location of the address. Again, the syntax resembles earlier examples. The only difference here is that we are using esri.geometry.Point:

```
//Display the address search results
function showResults(evt) {
  var geom;
  if(evt.addresses.length > 0)
  {
    geom = evt.addresses[0].location;
```

```
    if ( geom !== undefined ) {
        map.centerAndZoom(geom, 12);
   //add a graphic to the map at the geocoded location
    map.graphics.add(new esri.Graphic(
new esri.geometry.Point(evt.addresses[0].location.x, evt.addresses[0].
location.y, map.spatialReference),
new esri.symbol.SimpleMarkerSymbol(esri.symbol.SimpleMarkerSymbol.
STYLE_SQUARE, 10,
new esri.symbol.SimpleLineSymbol(esri.symbol.SimpleLineSymbol.STYLE_
SOLID,
new esri.Color([0,255,0]), 1),
new esri.Color([255,0,0,0.25]))
            ));
            }
        }
      }
```

You can find the code for this example at:

http://www.webmapsolutions.com/book/mobilearcgis/chapter4/
MobileMapFindAddress.html.

Before we finish with geocoding, it is worth mentioning the new Esri Geocoder widget. This greatly simplifies finding map locations. Read more at the following link:

https://developers.arcgis.com/javascript/jssamples/locator_simple.html.

Summary

In this chapter, we advanced further with the ArcGIS JavaScript API for mobile development. We looked at the code for four different tools that are often requested by mobile ArcGIS users. These included providing a feature details popup, layer legend, feature search, and address search. We got to use Dojo widgets for the first time. Esri provides many widgets that should greatly help your development efforts. Tools that can be customized are the cornerstone of many ArcGIS mobile apps. This chapter takes us nicely towards our next chapter: providing custom mobile apps that can be used on different devices. Often, users want to access their ArcGIS app on their smartphone, phablet, or tablet. Responsive design is a popular approach to developing web applications that work well on all devices. We will be covering this in the next chapter.

5

Providing Cross-device Support with Responsive Design

Often, mobile ArcGIS web applications are accessed from a variety of mobile devices: smartphones, phablets, and tablets. Using responsive design, a mobile web application can be built that looks good and performs well across all mobile devices. Again, by taking a hands-on and step-by-step approach, we will build a simple ArcGIS mobile responsive application.

In this chapter, we will cover the following topics:

- Approaches to cross-device support
- The magic of style sheets
- Responsive design using Bootstrap
- Application testing

Approaches to cross-device support

In *Chapter 2, Understanding Mobile Frameworks and APIs*, we made a brief mention of responsive design. This is an approach to web application development that provides true cross-device capabilities, allowing your mobile app to look good and be user-friendly on all mobile devices. Bootstrap, which is the focus of this chapter, is the most popular responsive framework. The following screenshot illustrates ArcGIS web applications provide true cross-device capabilities:

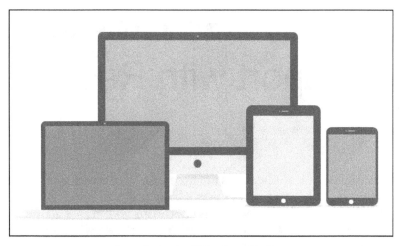

Cross-device ArcGIS web applications

Layouts built with Bootstrap adapt to different screen sizes, screen resolution/pixel density and changes in device orientation. As mentioned in *Chapter 2, Understanding Mobile Frameworks and APIs*, Bootstrap uses a **fluid grid system** that is based on percentages. The default grid system has 12 columns or a 940 pixel-wide container, as shown in the following screenshot:

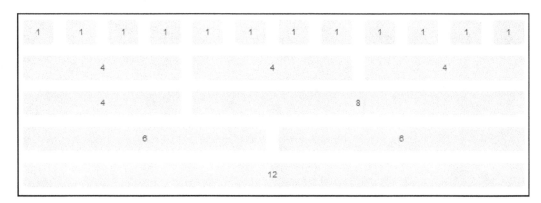

Default fluid grid system

Think of the grid as a page that is broken up into a series of columns. On a larger 10-inch tablet, 3 columns might be visible whereas the same page might show just 2 columns on a 6-inch phablet and, on a smartphone, it might show just a single column. As a device's screen size gets smaller, the screen content is stacked more. The following screenshot shows how the same page appears on a tablet and a smartphone. In this case, horizontal menu items on the tablet are stacked vertically on the smartphone:

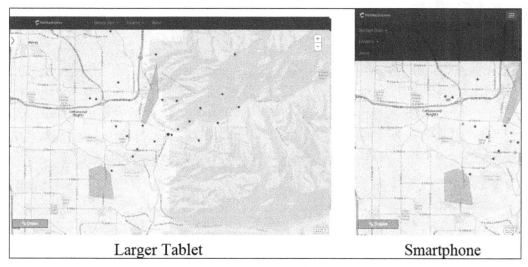

Responsive views

There are other ways to provide cross-device support in web applications. In addition to responsive design, **progressive enhancement** is another popular approach. We have already discussed jQuery Mobile in *Chapter 2, Understanding Mobile Frameworks and APIs*. This framework offers both responsive and progressive enhancement. Progressive enhancement is characterized by the following:

- Built with specific screen sizes in mind
- Formal control of UX
- Mimics native app look, feel, and behavior
- Multi-page views
- Suitable for use in Hybrid applications

The following screenshot shows a site built using progressive enhancement:

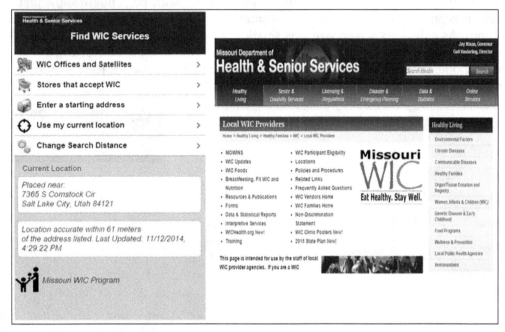

Progressive enhancement

Responsive and progressive enhancement differ in their delivery of responsive/adaptive structures. Responsive enhancement relies on flexible and fluid grids, whereas progressive enhancement relies on predefined screen sizes. Progressiveness takes a layered approach that utilizes scripting and CSS to assist in adapting to various devices and screen sizes.

The magic of style sheets

At the heart of both responsive and progressive approaches to cross-device development is **Cascaded Style Sheets** (**CSS**). This is the industry standard for styling the presentation and positioning of web page elements. Bootstrap relies on CSS Media queries that allow developers to specify when certain CSS rules should be applied. They can be used to deliver a tailored style sheet to a specific device. There are many places where we can add these queries to our code. For example, they can be links in our HTML:

```
<link rel="stylesheet" type="text/css"
  media="screen and (max-device-width: 480px)"
  href="responsiveArcGIS1.css" />
```

The media parameter in the preceding code contains two components: A media type (screen) represents the actual query enclosed within parentheses containing a particular media feature (max-device-width) to inspect, that is followed by the target value (480px)

What we are doing here is asking the device whether its horizontal resolution (max-device-width) is equal to or less than 480 pixels. If the test passes, that is, if we're able to view our work on a small-screen device like the iPhone 3, then the device loads responsive ArcGIS1.css. Otherwise, the link is ignored altogether. The queries can also include CSS as part of a @media rule:

```
@media screen and (max-device-width: 480px) { .. }
```

You will see variations of this CSS syntax, which are as follows:

```
/* Extra small devices (phones, less than 768px) */
/* No media query since this is the default in Bootstrap */

/* Small devices (tablets, 768px and up) */
@media (min-width: @screen-sm-min) { ... }

/* Medium devices (desktops, 992px and up) */
@media (min-width: @screen-md-min) { ... }

/* Large devices (large desktops, 1200px and up) */
@media (min-width: @screen-lg-min) { ... }
```

 You can read more about media queries in the official Bootstrap documentation at:
http://getbootstrap.com/css/#grid-media-queries.

Progressive enhancement in jQuery Mobile uses roles and themes. The role defines the container type such as page, header, and so on, and the theme controls specific visual elements such as fonts, colors, gradients, shadows, and corners. Modifying a theme allows you to control the visual elements of objects such as buttons. In the following piece of code, we have shown container elements within a page that contain a reference to both role and theme:

```
<div data-role="page" id="home">
  <div data-role="collapsible" data-collapsed="true" data-theme="a"/>
</div>
```

This is the corresponding CSS code snippet that styles these containers:

```
html,body, div[data-role ="page"] { .. }
.ui-bar-a { .. }
```

Progressive enhancement and jQuery Mobile, though not the focus of the book, are worth further investigation by readers. They provide a different approach to ArcGIS mobile development and cross-device support.

 The official jQuery Mobile website with complete documentation can be found at:

`http://jquerymobile.com/`.

Responsive design using Bootstrap

It's time to dig into Bootstrap. As we have done in previous chapters, let's start with a simple, clean code base. We will actually modify the code base that we started with in *Chapter 3, Building Your First ArcGIS Mobile Application*. First, we need to edit our viewport and modify our content attribute by adding *width=device-width*:

```
<meta name="viewport" content="width=device-width, initial-scale=1.0">
```

The special value, device-width, is the width of the screen in pixels at a scale of 100 percent in CSS.

We next need to add the compiled and minified Bootstrap CSS and JavaScript:

```
<link href="https://maxcdn.bootstrapcdn.com/bootstrap/3.3.1/css/
bootstrap.min.css" rel="stylesheet">
<script src="https://maxcdn.bootstrapcdn.com/bootstrap/3.3.1/js/
bootstrap.min.js"></script>
```

 The process of minification removes all unnecessary characters from the source code without changing its functionality. This makes the library size smaller which helps to improve download speeds and browser parsing performance. Our base code should now look as follows:

```
<html>
  <head>
    <meta name="viewport" content="width=device-width, initial-
scale=1.0">
<title>Simple Responsive Map</title>
    <!-- stylesheets -->
<link rel="stylesheet" href="https://maxcdn.bootstrapcdn.com/
bootstrap/3.3.1/css/bootstrap.min.css">
```

```html
<link rel="stylesheet" href="https://maxcdn.bootstrapcdn.com/
bootstrap/3.3.1/css/bootstrap-theme.min.css">
<link rel="stylesheet" href="http://js.arcgis.com/3.12/esri/css/esri.
css">
<!-- script -->
<script src="https://ajax.googleapis.com/ajax/libs/jquery/1.11.1/
jquery.min.js"></script>
<script src="https://maxcdn.bootstrapcdn.com/bootstrap/3.3.1/js/
bootstrap.min.js"></script>
<script src="http://js.arcgis.com/3.12compact/"></script>
    <style>
        html, body, #map {
          height: 100%;
          width: 100%;
        margin: 0;
        padding: 0;
          }
    </style>
     <script>
      require(["esri/map", "esri/layers/FeatureLayer", "dojo/parser",
"dojo/dom", "dojo/domReady!"], function(Map, FeatureLayer, parser,
dom) {

   parser.parse();
  //map object
        var map = new Map("map", {
    center: [-118, 34.5],
    zoom: 7,
    basemap: "streets"
        });
          var marathon = new FeatureLayer("http://services.
arcgis.com/V6ZHFr6zdgNZuVG0/arcgis/rest/services/Boston_Marathon/
FeatureServer/0",
      {
        mode: FeatureLayer.MODE_ONDEMAND,
          outFields:["*"]
      });
    map.addLayers([marathon]);
      });
        </script>
  </head>
  <body>
      <div id="map"></div>
  </body>
</html>
```

Notice that we have included two Dojo toolkit references here: **domparser, and dojo 'modules'**. These will be needed as we start to add tools. You can view this base code at:

```
http://webmapsolutions.com/book/mobilearcgis/chapter5/
MobileResponsiveMapBase.html
```

Let's now play with the page layout. This involves adding the <div> tags to the body of our web page.

 The <div> tag defines a division or section in an HTML document. It is used to group block elements to format them in CSS.

Adding responsive page elements

We will build the user interface of the responsive ArcGIS mobile application, as shown in the following screenshot:

ArcGIS application header

We will build an application with a header that contains a number of tools similar to what we built in *Chapter 4, Advancing the Basic Mobile ArcGIS Application*. At present we have a single <div> tag in the page body. We will add <div> and a <nav> tag to now include a header.

 The <nav> tag defines a set of navigation links. The <nav> element is intended only for major blocks of navigation links. For more information, see http://www.w3schools.com/tags/tag_nav.asp.

The HTML5 class attribute is used to apply Bootstrap classes on these page elements. We will start with the header or **navbar**:

```
<div class="bs-example">
    <nav id="myNavbar" class="navbar navbar-default"
role="navigation">
    </nav>
</div>
<!-- Map -->
<div id="map" ></div>
```

You will notice that the `<nav>` tag has a class attribute that references navbar. This is a Bootstrap component. As stated in the Bootstrap documentation: "Navbars are responsive meta components that serve as navigation headers for your application or site. They begin collapsed (and are toggleable) in mobile views and become horizontal as the available viewport width increases."

The `role="navigation"` attribute is needed to help with accessibility. HTML5 has a variety of section elements. The `<nav>` element is one of these elements. This helps users find the navigation sections of a page. Adding this navigation role helps user agents that don't support HTML5 understand the structure. To the navbar, we will add a logo and tools including linked geolocation text element, a drop-down menu, and a search input box and button. To do this, inside the `<nav>` tag, add the following code:

```
<nav id="myNavbar" class="navbar navbar-default navbar-fixed-top"
role="navigation">
  <div class="container">
        <div class="navbar-header">
    <!-- Logo -->
                    <img src="http://plagatux.es/wp-content/
uploads/2013/10/photo1.jpg"          alt="" class="navbar-brand"/>
        </div>
    <!-- Tools -->
        <div class="collapse navbar-collapse" >

                    </div>
    </div>
</nav>
```

The second `<div>` in the preceding code contains our tools. By using the `collapse nav-bar-collapse class`, we ensure that these link element tools are collapsible into their vertical mobile view when the viewport is narrow, such as on a smartphone. We will cover this topic further later. Our tools `<div>` should now look like the following:

```
<!-- Tools -->
<div class="collapse navbar-collapse" id="bs-example-navbar-
collapse-1">
    <ul class="nav navbar-nav">
        <li><a href="#">GeoLocate</a></li>
        <li class="dropdown">
    <a href="#" data-toggle="dropdown" class="dropdown-toggle">Tools
<b class="caret"></b></a>
        <ul class="dropdown-menu">
            <li><a href="#" >Legend</a></li>
        </ul>
        </li>
    </ul>
    <ul class="nav navbar-nav navbar-right">
      <form class="navbar-form navbar-left" role="search">
      <div class="form-group">
        <input id="searchText" type="text" class="form-control"
value="380 New York St, Redlands">
      </div>
      <a href="#" >Search</a>
      </form>
      </ul>
</div>
```

Notice that each of the HTML elements has a class attribute, for example `class='collapse navbar-collapse'`. Many of these are the Bootstrap classes which make the application responsive.

> You can read more about Bootstrap components in the official Bootstraps documents at:
>
> http://getbootstrap.com/components/.

Let's incorporate these additions into our core code base for this chapter. The image below shows how the application now looks.

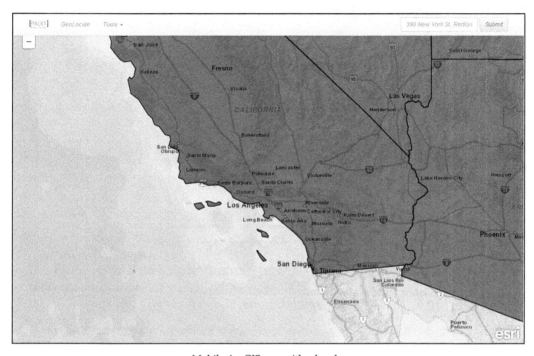

Mobile ArcGIS app with a header

This looks good. We have our header and all the UI elements in place.

You can see a live sample of this code at:

`http://webmapsolutions.com/book/mobilearcgis/`
`chapter5/MobileResponsiveMap1.html`.

Let's now resize this window. Place your mouse at the bottom right corner of the browser window, hold it down, and resize. This is a very simple way for us to see how the application looks on a smaller device.

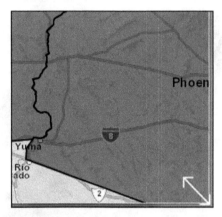

Resizing the browser window

You will notice that, as we resize the browser window below a certain size, our header tool options disappear. Ideally we would collapse the header tools into an accessible vertical view. To do this, let's add some code just above the logo `<img>` tag:

```
<button type="button" class="navbar-toggle" data-toggle="collapse"
data-target="#bs-example-navbar-collapse-1">
    <span class="sr-only">Toggle navigation</span>
    <span class="icon-bar"></span>
    <span class="icon-bar"></span>
    <span class="icon-bar"></span>
</button>
```

This code provides a button which can open and collapse our toolbar. When you resize the browser window, you should see the following two images repectively:

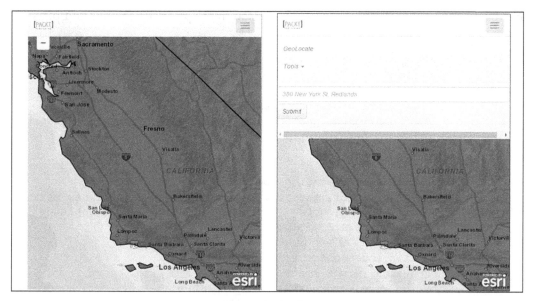

Responsive header

The left image shows what the application looks like on smaller mobile devices. This is the view when the toolbar is closed . The image on the right side shows the look when the header button is tapped and reveals the tools. Now you can see the power of Bootstrap. With the addition of simple classes to `<div>` tags, a Web application becomes responsive or usable across devices.

 Again, you can view this code sample at:
`http://webmapsolutions.com/book/mobilearcgis/`
`chapter5/MobileResponsiveMap2.html`

This is a good start. Next, let's add some functionality.

Responsive tools

As mentioned earlier, we will add some of the tools that we built in *Chapter 4, Advancing the Basic Mobile ArcGIS Application*, to this application. Let's first incorporate the geolocate option. To the geolocate page element, add an `onclick` function call:

```
<li><a href="#" onclick="geolocate()">GeoLocate</a></li>
```

Next, add the `geolocate` functions that we used in the previous chapter and place them in a new `<script>` tag:

```
<script>
    //GeoLocate
    function geolocate() {
        if (navigator.geolocation) {
            navigator.geolocation.getCurrentPosition(zoomToLocati
on, locationError);
        }
    }
    function zoomToLocation(location) {
        var pt = esri.geometry.geographicToWebMercator(new esri.
geometry.Point(location.coords.longitude, location.coords.latitude));
        map.centerAndZoom(pt, 16);
    }
    function locationError(error) {
        switch (error.code) {
            case error.PERMISSION_DENIED:
                alert("Location not provided");
                break;
            case error.POSITION_UNAVAILABLE:
                alert("Current location not available");
                break;
            case error.TIMEOUT:
                alert("Timeout");
                break;
            default:
                alert("unknown error");
                break;
        }
    }
    //end GeoLocate
</script>
```

Variable scoping is worth a brief discussion. Look at the sample code for this example at:

```
http://webmapsolutions.com/book/mobilearcgis/chapter5/
MobileResponsiveMap3.html.
```

We have made the map object global by creating a top level script element:

```
<script>
 var map;
 var locator;
</script>
```

We've also added `locator` for our geocode service, which will be discussed next. By making these global, we can reference the map object elsewhere in the code.

Next, we need to add the geocode service or address search. Again, we will use the code from the previous chapter for this tool. Below the `FeatureLayer` object in the code, add the following:

```
locator = new Locator("http://geocode.arcgis.com/arcgis/rest/services/
World/GeocodeServer");
locator.on("address-to-locations-complete", showResults);
```

A Locator object represents a geocode service resource exposed by the ArcGIS Server REST API. It is used to generate candidates for an address, and also to find an address for a given location.

Now, create a new `<script>` element and add the following piece of code:

```
<script>
//Address Search
//Execute find address request
  function findAddress(searchText) {
  //Clear any existing graphics
      var address = {
      "SingleLine": searchText
      };
      locator.outSpatialReference = map.spatialReference;
      var options = {
      address: address,
      outFields: ["Loc_name"]
      };
      locator.addressToLocations(options);
  }
  //Display the address search results
  function showResults(evt) {
    var geom;
    if(evt.addresses.length > 0)
    {
    map.graphics.clear();
      geom = evt.addresses[0].location;
      if ( geom !== undefined ) {
        map.centerAndZoom(geom, 12);
      }
    }
    }
  </script>
```

SingleLine in the address options above specifies the location to be geocoded. This can be a street address, place name, postal code or POI. In our case, it is the user's address input. The locator begins searching for matches when the addressToLocations method is called.

Finally, update the Submit button, add a function call with onclick, and pass the address or searchText to the findAddress function:

```
<button type="submit" class="btn btn-default"
onclick="findAddress(dojo.byId('searchText').value);">Submit</button>
```

You can see this code running at:

http://webmapsolutions.com/book/mobilearcgis/chapter5/
MobileResponsiveMap4.html

The map now zooms to the address. One element that we have left off here is adding a point graphic to the map to mark this location. We have left this for the readers to add. You can refer to the code in the previous chapter for assistance.

Before moving on and adding our final tool, let's make a couple of minor changes to our code. Firstly, you will notice that the zoom tool is partly obscured by the header:

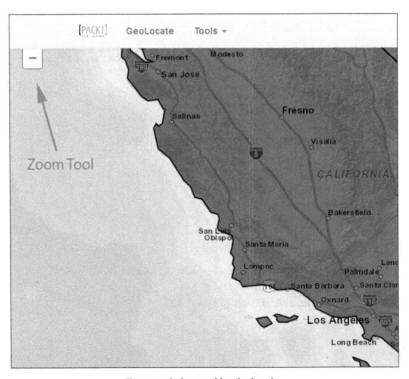

Zoom tool obscured by the header

Let's reposition this tool to remove this overlap by adding the following piece of code to the `<style>` element:

```
#map_zoom_slider{
     top:10% !important;
}
```

Now, let's make one other addition. If we select a State with a finger tap, no details popup appears. This is because we have not added our nice mobile-enabled popup from *Chapter 4, Advancing the Basic Mobile ArcGIS Application*. We encourage readers to try and add this to the current code base. For now, let's add the default feature details popup to our application. To do this, you need to add the ArcGIS `InfoTemplate` module. The code addition can be seen below:

```
var template = new InfoTemplate();
//Feature layer
 var marathon = new FeatureLayer("http://services.arcgis.
com/V6ZHFr6zdgNZuVG0/arcgis/rest/services/Boston_Marathon/
FeatureServer/0",
   {
    mode: FeatureLayer.MODE_ONDEMAND,
     outFields:["*"],
     infoTemplate: template
   });
```

The completed sample can be found at:

http://webmapsolutions.com/book/mobilearcgis/chapter5/
MobileResponsiveMap5.html.

The following screenshot shows how these code additions look on a smartphone:

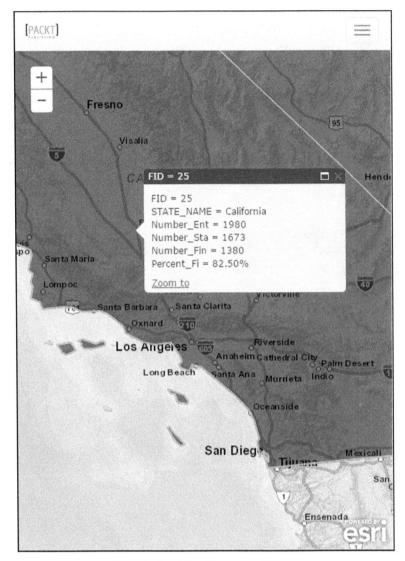

Zoom tool and details' popup additions

This feels a little more complete.

We have one more tool to add. Let's reuse some of our legend code from the previous chapter. However, instead of adding a side panel, let's make the legend so that it covers the full screen and allows us to switch between the map and map legend. This provides a better mobile experience. The following screenshot shows on the left, the application and on the right the legend view:

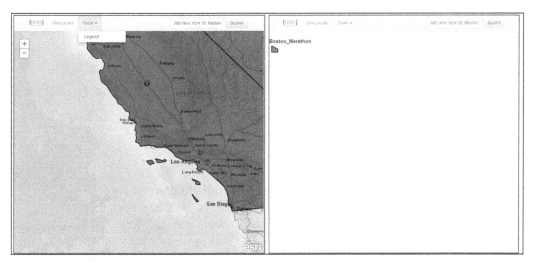

Responsive legend

The completed code sample is available at:

`http://webmapsolutions.com/book/mobilearcgis/chapter5/ MobileResponsiveMap6.html`.

Let's walk through all the code additions. We first need to add the legend widget to `require()`:

```
require(["esri/map", "esri/layers/FeatureLayer", "esri/tasks/locator",
"dojo/parser", "dojo/dom", "esri/InfoTemplate", "esri/dijit/Legend",
"dojo/domReady!"], function(Map, FeatureLayer, Locator, parser, dom,
InfoTemplate, Legend)
```

Next, let's add a `<div>` tag below our map tag:

```
<div id="legendDiv"></div>
```

We will be switching between map and legend views. So, when the page loads initially, we need to make the legend tab invisible. To do this, add the following piece of code at the end of `require()`:

```
document.getElementById("legendDiv").style.display = "none";
```

Let's next add the `legend` object. To do this, above the hide legend code, add the legend widget as follows:

```
//Legend
map.on("layers-add-result", function (evt) {
   var layerInfo = dojo.map(evt.layers, function (layer, index) {
   return {layer:layer.layer, title:layer.layer.name};
   });
   if (layerInfo.length > 0) {
   var legendDijit = new esri.dijit.Legend({
     map: map,
     layerInfos: layerInfo
   }, "legendDiv");
   legendDijit.startup();
   }
});
```

The legend dijit displays a label and symbol for some or all of the layers in the map. These are based on the symbol or renderer for each layer. The legend supports the following layer types: `ArcGISDynamicMapServiceLayer`, `ArcGISTiledMapServiceLayer`, `FeatureLayer`, `CSVLayer`, `KMLLayer`, and `WMSLayer`. The **layerInfos** option allows developers to control the layers displayed, the layer symbol and label.

Now, update the Legend `<li>` page element with an `onclick` attribute:

```
<li><a href="#"  onclick="showLegend(this)"; >Legend</a></li>
```

When we select legend from the drop-down header menu, we invoke the `showLegend` function. Add a new script element for this function:

```
<script>
//Legend show/hide
function showLegend(evt) {
   switch (document.getElementById("legendDiv").style.display) {
     case "none":
        document.getElementById("legendDiv").style.display = "block";
        document.getElementById("map").style.display = "none";
        evt.innerText = "Map";
        break;
```

```
    case "block":
        document.getElementById("legendDiv").style.display = "none";
        document.getElementById("map").style.display = "block";
        evt.innerText = "Legend";
        break;
    }
}
</script>
```

Here, each time you select Legend in the Tools menu, the Legend `<div>` becomes visible while the map `<div>` gets hidden. We also change the label from Legend to Map to help give a visual cue to users. Let's make one other small addition and move the legend icons so that they are below the header. To do this, we can use a similar style to the one that we used to move the zoom tool:

```
#legendDiv{
    top:10% !important;
}
```

`!important` in the above CSS sets the style to have the highest priority regardless of any other priority.

This is it. We have now created a responsive mobile ArcGIS application and added geolocation, address search, and legend tools.

It is important to test your application as you build it. In this chapter, we mentioned simple browser-window resizing to see how your responsive application looks on different screens. As discussed in *Chapter 3, Building Your First ArcGIS Mobile Application*, Chrome's DevTools also provides ways to emulate different mobile devices. As you move closer to a production ready version of your mobile ArcGIS application, real world cross-platform, cross-browser, cross-device testing on actual mobile devices will become important.

Summary

We covered a lot of ground in this chapter. Our goal was to extend what we had already learned and develop a responsive mobile application. Any mobile ArcGIS application should be usable across devices. While we chose to use the Bootstrap framework to provide this flexibility, jQuery and progressive enhancement are other good options that we encourage you to explore. Starting with a simple Bootstrap-enabled code sample, we moved our discussions forward from the previous chapter and built a responsive application that included geolocation, address search, and a legend.

This chapter introduced some more advanced concepts. This included adding Bootstrap responsive page elements and applying dojo widgets and modules respectively. In the next chapter, we will discuss new cloud-based data and services that can be leveraged by ArcGIS mobile web apps. ArcGIS Online is our next area of focus.

6
Integration with ArcGIS Online

This is an important chapter. Esri's official launch of ArcGIS Online in 2012 heralded a new approach to publishing, accessing, and purchasing ArcGIS services. Using a subscription-based model, ArcGIS Online has made the sharing and managing of geodata considerably easier. Mobile application developers need to understand ArcGIS Online. Authentication, users, groups and webmaps are key concepts. We will discuss these concepts in greater depth in this chapter, and illustrate by building a mobile ArcGIS Online application.

In this chapter we will cover the following:

- ArcGIS Server and ArcGIS Online.
- ArcGIS Online basics.
- Building an ArcGIS Online Mobile application.

Introduction

The cloud and mobile technology are having a profound affect on GIS. ArcGIS Online is Esri's first step into cloud-based GIS services. Pricing is subscription-based; a combination of annual fees and credits. ArcGIS Online is different. It has made publishing of both public and private data from different sources far easier. In many ways it is a big advancement, making ArcGIS services more affordable and accessible. Indeed cloud-based services like ArcGIS Online represent the future of GIS.

As the diagram below shows, ArcGIS Online provides a single access point for mobile and Web applications to access maps, map and feature layers, and GIS services.

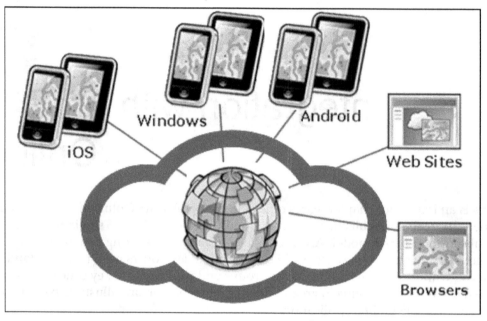

ArcGIS Online

ArcGIS Server and ArcGIS Online

ArcGIS Server has been at the core of Esri's product offerings for many years. It has provided GIS services to organizations across the world. As described by Esri:

 ArcGIS Server enables the publishing and sharing of GIS resources (e.g., maps, locators, GP models) as web services so they can be accessed and used over the Web/intranet. It is a product that you would typically install and configure on-premise in your organization.

In 2012 ArcGIS Online was launched, a cloud-based GIS "software as a service" or SaaS. Again from Esri:

"ArcGIS Online" is the online component of ArcGIS. It is a complete, cloud-based, collaborative content management system for working with geographic information. The platform provides an on-demand, secure, open, and configurable infrastructure for creating web maps, web-enabling your data, sharing your map, data, and applications, and managing content and multiple users from your organization.

ArcGIS Online provides a new more affordable way to publish and access geospatial data. Administrators can, through a simple interface, publish data from many different formats to ArcGIS Online for public or private consumption.

ArcGIS Online is at the heart of what Esri now describe as the ArcGIS platform. The ArcGIS platform offers two deployment models for web GIS. ArcGIS Online, is the cloud-based offering. This requires no hardware infrastructure for an organization to maintain because Esri manages and maintains ArcGIS Online. Alternatively, there is an on-premises model, which Esri describe as:

 The on-premises model includes Portal for ArcGIS. Portal for ArcGIS is deployed with ArcGIS for Server and optionally with the ArcGIS Data Appliance. In the on-premises model, an organization manages the hardware infrastructure to operate the ArcGIS platform.

This chapter is important because the concepts and coding we cover will be increasingly more relevant to mobile ArcGIS application developers over time.

ArcGIS Online basics

Before we cover the basics of ArcGIS Online, let's discuss your access to ArcGIS Online. Some readers will have employers with organizational accounts. From Esri:

 If you are not a member of an ArcGIS Online organization, you can create a public account to access ArcGIS Online. A public account is available for individuals to create and share content with limits on usage. If you already have an Esri Account, it is automatically a public account and you can use it to sign in.

That is one potential path. Another is to take advantage of Esri's free developer ArcGIS Online account. This is a limited usage account, which comes with 50 credits per month. It is an excellent way to take your first steps with ArcGIS Online.

Find out more about the ArcGIS Online developer account at:
`https://developers.arcgis.com/en/`

There are three key ArcGIS Online concepts which are important to cover: Named Users, web maps, and OAuth authentication.

Named Users and Groups

The concept of a Named User is important to ArcGIS Online. To quote Esri:

"Named User credentials give you a unique, secure identity in the ArcGIS platform. You can use the platform through permissions given to you by your administrator, join groups, and access resources that you own or have been shared with you. Basically, a Named User is a person licensed to use the software".

Notice the use of the term platform and not specifically ArcGIS Online. You will need to be a Named User to access ArcGIS Online but, most importantly, having an identity provides access to the overall ArcGIS platform.

Read more about named users in the Esri product documentation:

`http://www.esri.com/products/technology-topics/`
`named-user`

ArcGIS Online administrators have important controls over Named Users, which include access to data, grouping and much more.

Webmaps

Webmaps are at the heart of ArcGIS Online. Webmaps can be thought of as map mashups; the combination of a base map with layer overlays. Geodata published as layers in ArcGIS Online or ArcGIS Server can be combined and made available through a webmap. Other data sources include **comma-separated values (CSV)** files, GeoJSON files or zipped shapefiles, feature collections, ArcMap documents, ArcGIS Pro maps, or zipped file geodatabases. Esri provides a rich set of free base maps with ArcGIS Online.

ArcGIS Online administrators and publishers can create web maps. The process entails choosing a base map, adding appropriate layers, setting extent and access rights. Each web map, as we will see later, has a unique id.

Read more about webmaps in the Esri documentation:
`https://developers.arcgis.com/javascript/jshelp/`
`intro_webmap.html`

OAuth authentication

A key part of the code we will develop in the next section uses OAuth authentication. From Esri's documentation:

"OAuth 2.0 is an open standard for authorization over HTTP that allows apps to access server resources on behalf of an authorized user. It provides a mechanism that allows your user to authenticate without having to give their username and password directly to the your app. Instead, a successful login on a remote server, such as ArcGIS Online, will respond with an access token that your app can use to access protected resources on behalf of your user".

Building authentication into your mobile ArcGIS applications is important moving forward. This controls who you provide access to your data. As we will see, access to ArcGIS Online is controlled by OAuth authentication.

Again from Esri:

 OAuth user logins allow app users to authorize themselves on an ArcGIS portal and access secured resources for which they have permission, without the app handling the user's credentials. User logins can be implemented in all types of apps: browser-based Web apps, server-based Web apps, device or tablet-based apps, and desktop apps. A Web app redirects the user to the login page on the remote server, while device, tablet, and desktop-based apps use client side browser controls to integrate this login experience into the app. Your app receives a unique alphanumeric string known as an authorization code in response to a successful login. An authorization code provides proof to the server that the user has authenticated successfully and allows access to resources for which the user has permission.

Once authenticated, Named User can access webmaps, ArcGIS services and other ArcGIS Platform resources.

Read more about Oath authentication in the Esri developer documentation:

`https://developers.arcgis.com/authentication/`

Let's now show the relevance of these core concepts by jumping into code.

Building an ArcGIS Online mobile application

We will build in this next section a mobile ArcGIS application which has four elements: sign in, log in, maps list and webmap.

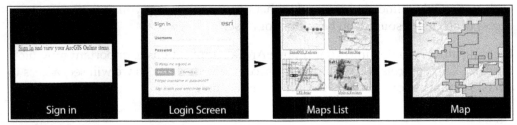

Application workflow

In this section, we will develop two separate applications. Authentication and map listing are in one application, and a webmap viewer in the other. Let's start with our code base for the first application.

```html
<html>
  <head>
    <meta name="viewport" content="width=device-width, initial-
scale=1.0">
  <title>ArcGIS Online Map</title>
      <!-- stylesheets -->
    <link rel="stylesheet" type="text/css" href="//js.arcgis.
com/3.11compact/dijit/themes/claro/claro.css">
    <link rel="stylesheet" href="http://js.arcgis.com/3.11/esri/css/
esri.css">
    <!-- script -->
  <script src="http://js.arcgis.com/3.11compact/"></script>
    <style>
        html, body, #map {
          height: 100%;
```

```
        width: 100%;
        margin: 0;
        padding: 0;
        }
    </style>
    <script>
    require(["dojo/dom-style", "dojo/dom-attr", "dojo/dom", "dojo/
on", "dojo/_base/array", "dojo/domReady!"], function(domStyle,
domAttr, dom, on, arrayUtils) {
    });
    </script>
  </head>
  <body class="claro">
  </body>
</html>
```

You will notice we have added to our require function some supporting Dojo modules "dojo/dom-style", "dojo/dom-attr", "dojo/dom", "dojo/on", "dojo/_base/array". These provide the key classes we need. A new stylesheet has also been added: claro.css. This is the newest theme for Dojo Dijit, and helps provide engaging visual styles and a professional look and feel to your mobile application.

> Read more about Dojo themes and theming in the official Dojo Docs:
>
> http://dojotoolkit.org/reference-guide/1.10/dijit/
> themes.html

As shown in the application workflow diagram above, this first page will actually display three screens. Let's add these to the `<body>` of our page:

```
    <!-- Sign In -->
    <div id="anonymousPanel" style="display: none; padding: 5px; text-
align: center;">
        <span id="sign-in" class="action">Sign In</span> and view your
ArcGIS Online items.
    </div>
    <!-- Log In -->
    <div id="personalizedPanel" style="display: none; padding: 5px;
text-align: center;">
```

```
      Welcome <span id="userId" style="font-weight: bold;"></span>
      -
      <span id="sign-out" class="action">Sign Out</span>
    </div>
    <!-- Map List -->
    <div id="itemGallery" class="esri-item-gallery" style="width:
100%;"></div>
```

This gives us our sign in, log in and map list UI elements. Next we need to add the core ArcGIS JavaScript modules to provide authentication functionality. These are `"esri/arcgis/Portal"`, `"esri/arcgis/OAuthInfo"`, and `"esri/IdentityManager"`. The Portal API allows application developers to work with users, groups, and content hosted in ArcGIS Online or Portal for ArcGIS. `OAuthInfo` provides information about our OAuth configuration. IdentityManager helps in managing user credentials.

Next we add the following code inside the require statements callback function, which instantiates the `OAuthInfo` class:

```
var info = new ArcGISOAuthInfo({
  //replace this appID
  appId: "q244Lb8gDRgWQ8hM",
  popup: false
});
esriId.registerOAuthInfos([info]);
```

The `appId` is important. To generate an `appID` for your version of this application, follow these steps (note, these steps and screenshots are current at the time of writing):

1. Sign in to your ArcGIS Online account.

2. Select the **My Contents** option at top of the page. As shown in the image below, provide the URL to your app, select the web mapping option, and give the application a title and descriptive tag. This URL is what you will use to test your application. There will be more on this when we get to App ID.

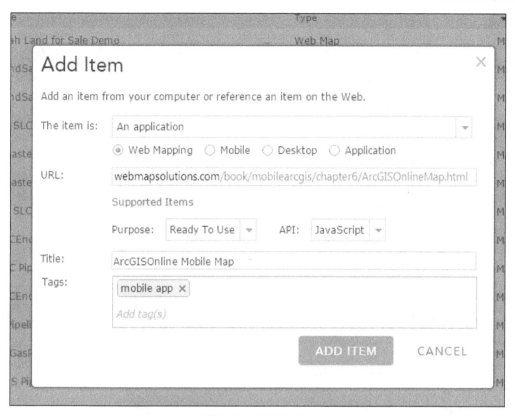

ArcGIS Online Add Application

3. Select **Add Item**.

4. In the page which loads select **Register**. Fill out the popup as shown below; the **App Type** is browser and use the URL you provided in step 3 above. When finished, click the **Add**, then **Register** button.

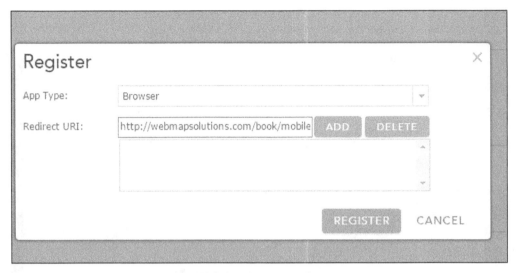

ArcGIS Online Register Application

In the page which comes up you will see the **App ID**:

ArcGIS Online App ID

This `App ID` is what you will need to add to your code.

Next let's control the `<div>` elements visibility. Add the following to the require statements `callback` function:

```
esriId.checkSignInStatus(info.portalUrl).then(
      function() {
         displayItems();
      }
   ).otherwise(
```

```
    function() {
        // Anonymous view
        domStyle.set("anonymousPanel", "display", "block");
        domStyle.set("personalizedPanel", "display", "none");
    }
);
//
function displayItems() {
}
```

This page controls the screens or page views we load. Though empty at the moment, the `displayItems` function forms part of the authentication process.

There are two more addition before we take a quick look at the application in a browser. First add the following CSS HTML `<style>` tag:

```
.action {
    color: blue;
    cursor: pointer;
    text-decoration: underline;
}
```

This provides our Sign In link. When you click on the **sign-in** button, it redirects you to the OAuth Log in page. Next, just above the `displayItems` function, add the following:

```
on(dom.byId("sign-in"), "click", function() {
    console.log("click", arguments);
    // user will be redirected to OAuth Log In page
    esriId.getCredential(info.portalUrl);
});
```

This redirects to the OAuth Log In page. Below are the **Sign In** and **Log In** screens.

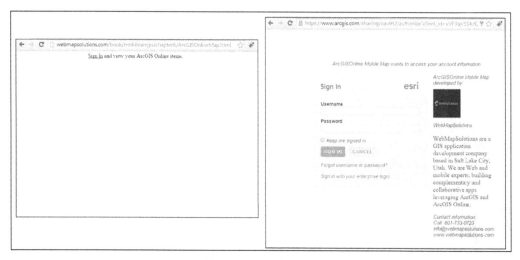

Sign In and Log in Pages

Right click to see the code in the following URL:

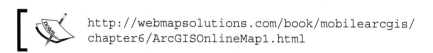

```
http://webmapsolutions.com/book/mobilearcgis/
chapter6/ArcGISOnlineMap1.html
```

Next we need to complete the `displayItems` function, and add a new function, `queryPortal`. The combination of these two functions authenticates the user based on their login information. As you will see in the code additions below, the `signIn()` process returns a promise which we capture through the `then` method.

```
function displayItems() {
 new arcgisPortal.Portal("https://www.arcgis.com").signIn().then(
   function(portalUser) {
      console.log("Signed in to the portal: ", portalUser);
      domAttr.set("userId", "innerHTML", portalUser.fullName);
      domStyle.set("anonymousPanel", "display", "none");
      domStyle.set("personalizedPanel", "display", "block");
      queryPortal(portalUser);
   }
 ).otherwise(
   function(error) {
      console.log("Error occurred while signing in: ", error);
   }
 );
}
```

```
//Query the portal
function queryPortal(portalUser) {
  var portal = portalUser.portal;
  var queryParams = {
    q: "owner:" + portalUser.username,
    sortField: "numViews",
    sortOrder: "desc",
    num: 20
  };
  portal.queryItems(queryParams).then(createGallery);

}
```

The `portal.queryItems` line above executes a query on the Portal. This returns a promise that, when resolved, returns `PortalQueryResult`. This contain a results array of `PortalItem` objects that match the input query. Using the `portalUser` object from the `OAuthInfo` class, we authenticate the user with ArcGIS Portal. If authentication is successful we call the `createGallery()` method. This function, shown below, simply takes the JSON object passed back from the ArcGIS Portal and displays the list of available web maps:

```
function createGallery(items) {
  var htmlFragment = "";
  arrayUtils.forEach(items.results, function(item) {
    htmlFragment += (
      "<div class=\"esri-item-container\">" +
        (
    item.thumbnailUrl ?
    "<div class=\"esri-image\" style=\"background-image:url(" + item.
thumbnailUrl + ");\"></div>" :
    "<div class=\"esri-image esri-null-image\">Thumbnail not
available</div>"
        ) +
        (
    item.title ?
    "<div class=\"esri-title\">" + (item.title || "") + "</div>" :
    "<div class=\"esri-title esri-null-title\">Title not available</
div>"
        ) +
      "</div>"
    );
  });
  dom.byId("itemGallery").innerHTML = htmlFragment;
}
```

Let's see what our application looks like.

Map List Page

 Notice that our application screenshots all have the same URL:

`http://webmapsolutions.com/book/mobilearcgis/`
`chapter6/ArcGISOnlineMap.html`

This is because, to test, we have to use the same URL we set with **appID**.

Right click to see this code at the following URL:

 `http://webmapsolutions.com/book/mobilearcgis/`
`chapter6/ArcGISOnlineMap2.html`

This is not a very attractive or useful page. Let's do some styling: add the following CSS within the `<style>` section.:

```
.esri-item-gallery .esri-item-container {
  float: left;
  text-align: center;
  padding: 10px;
  width: 204px;
  display: inline-block;
}
```

```css
.esri-item-gallery .esri-image {
  width: 200px;
  height: 133px;
  border: 2px solid gray;
  border-radius: 5px;
}
.esri-item-gallery .esri-null-image {
  line-height: 133px;
  text-align: center;
  color: #999999;
}
.esri-item-gallery .esri-title {
  white-space: nowrap;
  overflow: hidden;
  text-overflow: ellipsis;
}
.esri-item-gallery .esri-null-title {
  color: #999999;
}
```

We have a **Sign Out** link at the top of the page, so let's add some functionality after the on(dom.byId("sign-in") function, add the following:

```javascript
on(dom.byId("sign-out"), "click", function() {
  esriId.destroyCredentials();
  window.location.reload();
});
```

When a user selects **Sign Out** this functions destroys all credentials, and executes a page reload.

The map list page should now look as follows:

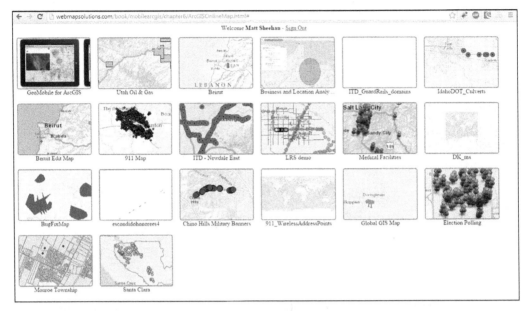

Map List Page after Styling

That is much better. Right click to see the code at the following URL:

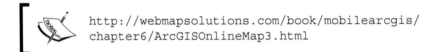

```
http://webmapsolutions.com/book/mobilearcgis/
chapter6/ArcGISOnlineMap3.html
```

We are moving forward nicely. What would be good now would be to select a map in the map list page and have that open in a separate map viewer. That means we will need to make one minor tweak to our existing code, and add a new web page which we will call WebMap.html. This will serve as our map viewer. Replace the code in line 144 with a hyperlink to WebMap.html and pass the webmap id as a parameter. To do this, add the following:

```
"<a href=\"WebMap.html?id=" + item.id + "\" target=\"_blank\"
class=\"esri-title\">" + (item.title || "") + "</a>" :
```

Right click to see the code at the following URL:

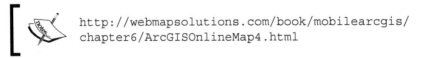

```
http://webmapsolutions.com/book/mobilearcgis/
chapter6/ArcGISOnlineMap4.html
```

Let's now write `WebMap.html`. This is a very simple page, shown below:

```html
<html>
  <head>
    <meta name="viewport" content="width=device-width, initial-
scale=1.0">
    <title>Simple WebMap</title>
        <!-- stylesheets -->
        <link rel="stylesheet" href="http://js.arcgis.com/3.11/esri/css/
esri.css">
        <!-- script -->
    <script src="http://js.arcgis.com/3.11compact/"></script>
        <style>
            html, body, #map {
                height: 100%;
                width: 100%;
                margin: 0;
                padding: 0;
                }
        </style>
        <script>
        require(["esri/map", "dojo/ready", "esri/arcgis/utils", "dojo/
domReady!"], function(Map, ready, arcgisUtils) {
            ready(function(){
            //Get the url object passed to the page
            var urlObject = esri.urlToObject(document.location.href);
             //If the url object has an webmapid. Create a map using this
id
            if(urlObject.query.id)
            {
                arcgisUtils.createMap(urlObject.query.id,"map").
then(function(response){
                    var map = response.map;
                    });
            }
            });
            });
        </script>
  </head>
  <body>
      <div id="map"></div>
  </body>
</html>
```

In the above snippet, very little coding has done a great deal. A web map is displayed in our application using the `esri.arcgis.utils` utility methods. The `esri.arcgis.utils.createMap` method creates a map using information from an `ArcGIS.com` web map. This information can be either a web map id or a by value representation of the web map. Right click to see the page source at the following URL:

`http://webmapsolutions.com/book/mobilearcgis/chapter6/WebMap.html`

You can load the full application by logging in using your ArcGIS Online credentials at the following URL:

`http://webmapsolutions.com/book/mobilearcgis/chapter6/ArcGISOnlineMap.html`

Summary

We have, in this chapter, introduced ArcGIS Online and some of the key concepts behind this new ArcGIS Platform service. In code, we examined authenticating a Named User in ArcGIS Portal, provided a list of webmaps available to the authenticated user, and a mechanism to load any of these webmaps in a map viewer. Understanding these concepts and the code we used in this chapter is important, particularly as you continue your ArcGIS mobile development journey.

So far we have remained focused on browser-based mobile development. One of the many advantages of web technology is that we are not only limited to browsers. There might be a need to build a mobile app which can be distributed in the various mobile app stores. Using technologies like PhoneGap we can produce a distributable ArcGIS app built with the ArcGIS JavaScript API. This will be the focus of our next chapter.

7
Developing Hybrid ArcGIS Mobile Applications with PhoneGap

Hybrid applications combine the elements of both Web and native applications. They are installed apps, which can be distributed via the various mobile app stores, Able to access the resources and sensors on mobile devices, these applications in many ways combine the best of all worlds. As we will discuss in this chapter using technologies like PhoneGap, a Web application built with the rich ArcGIS JavaScript API, can be turned into a native-like app which runs on the most popular mobile platforms.

In this chapter, we will cover the following topics:

- Introducing PhoneGap
- The PhoneGap setup
- Creating a test build
- Generating an Android certificate
- Developing hybrid ArcGIS mobile applications
- Additional code examples
- Plugins
- PhoneGap Build

Introducing PhoneGap

So far, we have discussed building mobile ArcGIS web apps. These are application which run in a browser. Let's imagine we want to build a mobile app which we can distribute in the various mobile app stores. An app which has access to the various sensors on a mobile device. A single code base which will run on all major mobile platforms: Android, iOS, Windows and Blackberry. Welcome to PhoneGap.

PhoneGap extends the features of HTML and JavaScript to work on a mobile device. Applications developed with PhoneGap are called **hybrid**. They are neither native (which rely on a platform's native UI framework) nor web-based. All layout rendering is done using web views, with the app having access to the native device APIs.

Nitobi developed PhoneGap which was purchased by Adobe in 2011. Cordova is the open source project that is the basis of PhoneGap. Though the names are somewhat interchangeable, PhoneGap is the trademarked name of the product that is today used by Adobe, mostly for commercial purposes. The following screenshot illustrates how the features of HTML, CSS and JavaScript are extended by PhoneGap:

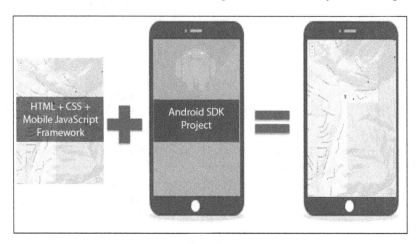

PhoneGap

Some of the core features of PhoneGap include:

- Native wrapper around a headless browser
- Cross-device access to device storage and sensors
- Create JavaScript apps for Google Play, App Store, and Windows Store
- Typically used in conjunction with jQuery

Mobile app Stores

PhoneGap has three basic components, which are as follows:

- **index.html**: This is the web page itself, that implements the application. This references the CSS, JavaScript, images, and other application resources
- **WebView**: The mobile app executes as a WebView within a native application wrapper
- **config.xml**: This provides the information about the app and specifies parameters that affect how it works

PhoneGap provides access to the device resources these include the camera, SD card, compass, accelerometer, connection state, contacts, and battery status.

As you will see, development of a hybrid app using PhoneGap follows a similar path to the development of a web application. However, we need to use an emulator to run our application. We use the command line to build the application and ultimately, produce a platform-specific package. We will use this for testing on a device. This package is what will ultimately be deployed to the app stores. The following screenshot show on the left an installed hybrid ArcGIS app icon, and on the right the app itself:

Device deployment

 The Cordova documentation can be found at `http://cordova.apache.org/docs/en/4.0.0/`.

Now it's time to setup our PhoneGap development environment.

PhoneGap setup

There are a number of ways to create a mobile app with PhoneGap. We will be using the Cordova command-line-interface or CLI. The CLI provides cross-platform support, meaning that we can generate hybrid apps for all supported platforms. Our focus in this chapter is to build and deploy an Android-specific mobile ArcGIS app.

We will spend some time in this section walking through the initial Cordova setup. This is a little time consuming, but it needs to be done only once. The following instructions walk you through a Windows setup:

1. First, install node.js. Go to `http://nodejs.org/` and select the Install button that will download the .msi file. Double-click and proceed with the setup.

2. Ensure that you have installed both a Java **SDK (Software Development Kit)** and the **JRE (Java Runtime Environment)**. Check that your **JAVA** and **JAVA_HOME** variables are correctly set.

> The **PATH** variable needs to contain the path to the bin folder of the Java installation, which is `C:\Program Files\Java\jdk1.8.0_05\bin`.
>
> The **JAVA_HOME** variable needs to be set to the root of that Java installation folder, which is `C:\Program Files\Java\jdk1.8.0_05` in our case.
>
> JAVA needs to be set to the Java installation bin directory, which is `C:\Program Files\Java\jdk1.8.0_05\bin` in our case.

3. Next, create a Cordova project with the Ripple emulator, as follows:

    1. Create a new directory, `c:\Cordova_ripple`.

    2. Open a command prompt and cd (change directory) to `c:\Cordova_ripple`.

    3. In command window type:

        ```
        >npm config set registry http://registry.npmjs.org/
        ```

    4. Install Cordova by the following command:

        ```
        >npm install cordova
        ```

> See a video here which walks through this part of the process: `https://www.youtube.com/watch?v=2Oh0gwlAeEg`

5. Add a new variable to your **PATH** (in Windows Explorer, right-click on **My Computer** and navigate to **properties | Advanced System Settings | Environment Variables**, select **PATH** and, click on **edit**, and add the following: `c:\Cordova_ripple\node_modules\.bin;%appdata%\npm\`

 Setting environment variables takes effect only for just command line windows. You should reopen a command prompt after adding new variables.

6. Next, install the Android SDK tools from `https://developer.android.com/sdk/installing/index.html`. Choose the Stand-alone SDK Tools. This is an installer. Once downloaded, click `installer_r24.0.2-windows.exe` to start the install process.

7. Note where the location where the installer saves the SDK on your system. In my case:

   `>C:\adt-bundle-windows-x86_64-20140702\sdk\tools\android`

   i) Add to your PATH environment variable the following:

   `C:\Users\$USER$\AppData\Local\Android\android-sdk\tools;`
   `C:\Users\$USER$\AppData\Local\Android\android-sdk\platform-tools`

   ii) Create a new environment variable called ANDROID_HOME with the path to your android-sdk:

   `C:\Users\$USER$\AppData\Local\Android\android-sdk`

8. When the installer is finished you will see a popup with options to install Android. If you have not already done it, you will need to install Android 4.4.2 (API 19). Scroll down the page and select the check-box. The install process can often take time.

9. Next, install the ripple-emulator by typing the following command:

   `>npm install -g ripple-emulator`

10. Then create a new Cordova project by typing the following command:

    `>cordova create foo`

11. Change directory into your new project (>cd foo) and add Android as a platform, which will generate the project files and directories:

    `>cordova platform add android`

    (there will be a lot of output...)

Under your foo project, you should now see the following files and folders:

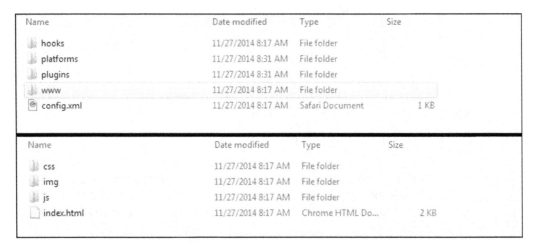

Cordova project directories

12. Finally, run the ripple command:

```
>ripple emulate --path platforms/android/assets/www
```

You should now see the following page load in your Chrome browser:

Ripple emulator

When we are developing our application, we will be working in our foo www directory (c:\Cordova_ripple\foo\www). The contents of this directory are copied to platforms/android/assets/www. As you can see in the preceding command, this is the path that we use for the ripple emulator.

Under the www directory you will notice that an index.html file has been generated. This is the page that we saw in the ripple emulator in the preceding image. Before we jump into ArcGIS, let's build this page and test it on an actual device.

Test Build

To build our project, we need to run the following command:

```
>cordova build --release android
```

 Note that you will need to make sure that you have installed WinAnt, which is available at https://code.google.com/p/winant/. You might need to open a new command prompt after the installation.

You will see a lot of output that will end with the following data:

```
-post-package:

-release-prompt-for-password:

-release-nosign:
     [echo] No key.store and key.alias properties found in build.properties.
     [echo] Please sign c:\Cordova_ripple\foo\platforms\android\ant-build\Cordov
aApp-release-unsigned.apk manually
     [echo] and run zipalign from the Android SDK tools.
[propertyfile] Updating property file: c:\Cordova_ripple\foo\platforms\android\a
nt-build\build.prop
[propertyfile] Updating property file: c:\Cordova_ripple\foo\platforms\android\a
nt-build\build.prop
[propertyfile] Updating property file: c:\Cordova_ripple\foo\platforms\android\a
nt-build\build.prop
[propertyfile] Updating property file: c:\Cordova_ripple\foo\platforms\android\a
nt-build\build.prop

-release-sign:

-post-build:
     [move] Moving 1 file to c:\Cordova_ripple\foo\platforms\android\ant-build
     [move] Moving 1 file to c:\Cordova_ripple\foo\platforms\android\CordovaLib\
ant-build

release:

BUILD SUCCESSFUL
Total time: 11 seconds
Built the following apk(s):
     c:\Cordova_ripple\foo\platforms\android\ant-build\CordovaApp-release-unsigne
d.apk

c:\Cordova_ripple\foo>
```

Android package build

You should see **BUILD SUCCESSFUL** and the path to a generated **apk** (**Android application package**),

```
C:\Cordova_ripple\foo\platforms\android\ant-build\CordovaApp-release-
unsigned.apk.
```

This is an unsigned version of your application.

> Android application package (APK) is the package file format used to distribute and install Android application software. This is the format required by Google Play.

Before we go ahead and test this apk on an Android device, we need to generate a certificate so that we can sign our apk.

Generating an Android certificate

To generate an Android certificate, you need two commands in your PATH environment variable: `keytool` and `jarsigner` (in our case, both are in `c:\Program Files\Java\jdk1.8.0_25\bin`).

First, run the following command:

```
>keytool -genkey -v -keystore my-release-key.keystore -alias alias_name
-keyalg RSA -keysize 2048 -validity 10000
```

You will be asked a number of questions and you need to complete each question. Most importantly you will be asked for a password. Set this to be something that is easy to remember. (There are actually two passwords that are requested, one at the start and one at the end. Set these passwords to be the same.) If you look now in your current directory, you will see a new file called `phonegapkey.keystore`.

Next, make sure that `CordovaApp-release-unsigned.apk` is in your current directory with the `phonegapkey.keystore file`. Now, run the following command:

```
>jarsigner -verbose -sigalg SHA1withRSA -digestalg SHA1 -keystore my-
release- key.keystore CordovaApp-release-unsigned.apk alias_name
```

`CordovaApp-release-unsigned.apk` should now be signed and installable on your mobile application. A simple way to copy this apk to your device is to email it to yourself. On your Android smartphone or tablet, ensure that you have checked Unknown sources (by navigating to **Settings** | **Security** | **Unknown sources**), as this allows apps other than those in Google Play to be installed.

For all the details on app signing, see the Android document at `http://developer.android.com/tools/publishing/app-signing.html`.

Open the email on your Android device, select the attached apk, and follow the installation instructions. After installing and opening the app, you should see the Apache Cordova device is ready:

Android app on a device

Though we are only covering Android in this chapter, as we mentioned earlier, PhoneGap/Cordova supports many other platforms that include Windows, iOS, and BlackBerry.

 See the Cordova documentation at `http://cordova.apache.org/docs/en/3.1.0/` for more information on developing mobile apps for the different platforms.

We have spent some time discussing the PhoneGap setup. The good news is that this needs to be done only once and development from here should move forward smoothly. It's finally time to jump into code.

Developing hybrid ArcGIS mobile applications

Let's start by looking at our pregenerated code in `index.html` in the www directory of our foo project:

```html
<html>
    <head>
        <meta charset="utf-8" />
        <meta name="format-detection" content="telephone=no" />
        <meta name="msapplication-tap-highlight" content="no" />
        <!-- WARNING: for iOS 7, remove the width=device-width and
height=device-height attributes. See https://issues.apache.org/jira/
browse/CB-4323 -->
        <meta name="viewport" content="user-scalable=no, initial-
scale=1, maximum-scale=1, minimum-scale=1, width=device-width,
height=device-height, target-densitydpi=device-dpi" />
        <link rel="stylesheet" type="text/css" href="css/index.css" />
        <title>Hello World</title>
    </head>
    <body>
        <div class="app">
            <h1>Apache Cordova</h1>
            <div id="deviceready" class="blink">
                <p class="event listening">Connecting to Device</p>
                <p class="event received">Device is Ready</p>
            </div>
        </div>
        <script type="text/javascript" src="cordova.js"></script>
        <script type="text/javascript" src="js/index.js"></script>
    </body>
</html>
```

As you can see, this is very simple code. We see a reference to `css/index.css`, `cordova.js`, and `js/index.js`.

Let's now add a new file to our www directory and call it `indexmap.html`. Next, at the top level of our `foo` directory (`C:\Cordova_ripple\foo`), you will see a file named `config.xml`. Open this file and set `indexmap.html` as the page where you want the application to launch:

```xml
<content src="indexmap.html" />
```

Below is our new `indexmap.html` file:

```html
<html>
<head>
    <meta http-equiv="Content-Type" content="text/html;
charset=utf-8">
    <meta name="viewport" content="width=device-width, initial-
scale=1, maximum-scale=1, user-scalable=no"/>
    <title>Basic Map</title>
    <style  type="text/css">
        body, html {
            width: 100%;
            height: 100%;
            margin: 0;
            padding: 0;
        }
        #mapDiv {
            height: 100%;
            width: 100%;
            margin: 0px;
        }
    </style>
</head>
<body>
    <div id="mapDiv"></div>

    <!-- Load the library and CSS references for ArcGIS API for
JavaScript -->
    <link rel="stylesheet" type="text/css" href="http://js.arcgis.
com/3.8/js/esri/css/esri.css">
    <script src="http://js.arcgis.com/3.8compact"></script>
    <script type="text/javascript" src="cordova.js"></script>
    <script type="text/javascript">
        var app = {
            // Application Constructor
            initialize: function() {
                this.bindEvents();
            },
            // Bind Event Listeners
            //
            // Bind any events that are required on startup. Common
events are:
            // 'load', 'deviceready', 'offline', and 'online'.
            bindEvents: function() {
```

```
                    document.addEventListener('deviceready', this.
        onDeviceReady, false);
                },
                // deviceready Event Handler
                //
                // The scope of 'this' is the event. In order to call the
        'receivedEvent'
                // function, we must explicity call 'app.
        receivedEvent(...);'
                onDeviceReady: function() {
                    app.receivedEvent('deviceready');
                },
                // Update DOM on a Received Event
                receivedEvent: function(id) {
                    console.log('Cordova device ready event: ' + id);

                    require(["esri/map"],
                        function(Map) {
                            // Create map
                            var map = new Map("mapDiv",{
                                basemap: "satellite",
                                center: [-122.69, 45.52],
                                zoom: 3
                            });
                        }
                    );
                }
            };
            app.initialize();
        </script>
    </body>
    </html>
```

You will notice that we have the ArcGIS style sheet and library references that we used in earlier chapters. Overall, this code sample looks a little different to our earlier examples. Here, we have an initialization function that is called using app. initialize() and we are making use of binding. In simple terms, we have set up a listener and functions for a deviceready event. Ultimately, we load a satellite base map when the device is ready. After building this code, re-signing it, and pushing it to our device, we should see the image in the following screenshot:

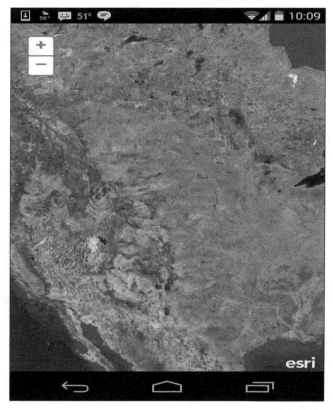

ArcGIS Android app on a device

There you have it! Your first installed hybrid mobile ArcGIS application.

Additional code examples

We have provided additional code examples based on the preceding code example that show GPS, a web map, and jQuery respectively. These code examples can be found at `http://webmapsolutions.com/book/mobilearcgis/chapter7/`.

Note that the provided examples will not actually render in a browser. After you load the page in your browser, right-click and select the source code. Copy this code to your own local version and build it as we did in the preceding code.

Andy Gup, an Esri technical evangelist, has provided some excellent ArcGIS resources in GitHub at `https://github.com/Esri/quickstart-map-phonegap`.

We have leveraged some of Andy's code samples as our additional examples.

Before we finish this chapter, let's briefly discuss plugins. They provide advanced features and simpler ways to compile our final hybrid mobile ArcGIS app for distribution.

Plugins

For an app to communicate with various device-level features, you need to add plugins that provide access to core PhoneGap/Cordova APIs. These include geolocation, network information, battery status, camera, and device orientation. A plugin is add-on code that provides an interface for native components.

For a full list of the 500 plugins, go to `http://plugins.cordova.io/#/viewAll`.

A plugin is added using the command line. You can add the geolocation by using the following command:

```
>cordova plugin add org.apache.cordova.geolocation
```

PhoneGap Build

Adobe provides a cloud-based service called PhoneGap Build that compiles PhoneGap applications. As we have seen, creating mobile applications is easy with PhoneGap, but there are challenges. PhoneGap Build helps to keep things simple.

PhoneGap Build can be found at `https://build.phonegap.com/`.

Among other things, PhoneGap Build helps to manage the compilation and signing of applications that are targeted at the various mobile platforms.

Summary

In this chapter, we introduced hybrid mobile applications. This underlines the true flexibility of JavaScript ArcGIS applications. Not only can we build cross-platform, cross-device ArcGIS applications that run in a browser, we can also build installable apps that can be distributed to the various app stores. The bulk of the chapter was spent walking through the setup rather than coding, since setup can be very confusing and frustrating. By spending additional time here, we hoped to help readers through this maze. After the setup, we provided a simple ArcGIS code example. Additional samples were also provided that showed different functionality. In this chapter we really only touched the surface of ArcGIS hybrid application development. We encourage readers to take what they have learned in earlier chapters and venture further into the world of hybrid ArcGIS app development with PhoneGap/Cordova.

Index

D

data input 12
declarative approach 30
Document Object Model (DOM) 28, 42
Dojo
  about 26-28
  plugins 29-31
  URL 28
  using, with ArcGIS API for
    JavaScript 33, 34
  widgets 29-31
Dojo themes
  URL 107

E

ECMAScript 20
enableHighAccuracy property 55
Esri product documentation
  URL 104
event object 53

F

fast responding mobile applications 19
feature popups 61-66
features
  finding 70-73
fluid grid system 35, 80

G

Geocoding service
  URL 74
Geographic Information Systems (GIS) 7
Geolocation API 35-37
getCurrentPosition() method 37
groups 104

H

hybrid 120
hybrid ArcGIS mobile applications
  code examples 131
  developing 128-130

I

interaction feedback
  providing 12
Internet Information Server (IIS) 44
iOS
  URL 16

J

jarsigner 126
Java Runtime Environment (JRE) 122
JavaScript 20
JavaScript development tools
  about 40-43
  browsers 43
  WebKit 43
JavaScript Geolocation API 54-56
JavaScript Sandbox
  URL 51
JavaScript Web Optimizer
  URL 45
jQuery Mobile
  about 21, 26, 31-33
  URL 84
jQuery sample
  URL 33
JS API, Esri
  URL 29

L

legend
  adding 66-70

M

map.centerAndZoom() function 56
map events
  listening for 51-54
map projections
  URL 55
Mashable
  URL 19
media queries
  URL 83

V

Vector Markup Language (VML) 34
viewport 42

W

watchPosition() function 56
WebKit 43
webmaps
  about 104
  URL 105
web mobile ArcGIS applications 20
web server
  setup 44
WinAnt
  URL 125

Thank you for buying
Developing Mobile Web ArcGIS Applications

About Packt Publishing

Packt, pronounced 'packed', published its first book, *Mastering phpMyAdmin for Effective MySQL Management*, in April 2004, and subsequently continued to specialize in publishing highly focused books on specific technologies and solutions.

Our books and publications share the experiences of your fellow IT professionals in adapting and customizing today's systems, applications, and frameworks. Our solution-based books give you the knowledge and power to customize the software and technologies you're using to get the job done. Packt books are more specific and less general than the IT books you have seen in the past. Our unique business model allows us to bring you more focused information, giving you more of what you need to know, and less of what you don't.

Packt is a modern yet unique publishing company that focuses on producing quality, cutting-edge books for communities of developers, administrators, and newbies alike. For more information, please visit our website at www.packtpub.com.

About Packt Open Source

In 2010, Packt launched two new brands, Packt Open Source and Packt Enterprise, in order to continue its focus on specialization. This book is part of the Packt Open Source brand, home to books published on software built around open source licenses, and offering information to anybody from advanced developers to budding web designers. The Open Source brand also runs Packt's Open Source Royalty Scheme, by which Packt gives a royalty to each open source project about whose software a book is sold.

Writing for Packt

We welcome all inquiries from people who are interested in authoring. Book proposals should be sent to author@packtpub.com. If your book idea is still at an early stage and you would like to discuss it first before writing a formal book proposal, then please contact us; one of our commissioning editors will get in touch with you.

We're not just looking for published authors; if you have strong technical skills but no writing experience, our experienced editors can help you develop a writing career, or simply get some additional reward for your expertise.

Administering ArcGIS for Server

ISBN: 978-1-78217-736-4   Paperback: 246 pages

Installing and configuring ArcGIS for Server to publish, optimize, and secure GIS services

1. Configure ArcGIS for Server to achieve maximum performance and response time.

2. Understand the product mechanics to build up good troubleshooting skills.

3. Filled with practical exercises, examples, and code snippets to help facilitate your learning.

Learning ArcGIS Geodatabases

ISBN: 978-1-78398-864-8   Paperback: 158 pages

An all-in-one start up kit to author, manage, and administer ArcGIS geodatabases

1. Covers the basics of building Geodatabases, using ArcGIS, from scratch.

2. Model the Geodatabase to an optimal state using the various optimization techniques.

3. Packed with real-world examples showcasing ArcGIS Geodatabase to build mapping applications in web, desktop, and mobile.

Please check **www.PacktPub.com** for information on our titles

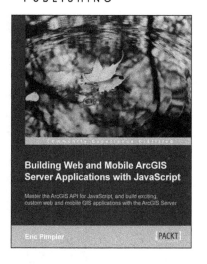

Building Web and Mobile ArcGIS Server Applications with JavaScript

ISBN: 978-1-84969-796-5   Paperback: 274 pages

Master the ArcGIS API for JavaScript, and build exciting, custom web and mobile GIS applications with the ArcGIS Server

1. Develop ArcGIS Server applications with JavaScript, both for traditional web browsers as well as the mobile platform.

2. Acquire in-demand GIS skills sought by many employers.

Programming ArcGIS 10.1 with Python Cookbook

ISBN: 978-1-84969-444-5   Paperback: 304 pages

Over 75 recipes to help you automate geoprocessing tasks, create solutions, and solve problems for ArcGIS with Python

1. Learn how to create geoprocessing scripts with ArcPy.

2. Customize and modify ArcGIS with Python.

3. Create time-saving tools and scripts for ArcGIS.

Please check **www.PacktPub.com** for information on our titles